T0254108

SpringerBriefs in Electrical and Computer Engineering

More information about this series at http://www.springer.com/series/10059

Zubair Md. Fadlullah • Nei Kato

Evolution of Smart Grids

Zubair Md. Fadlullah
Tohoku University
Sendai, Miyagi, Japan

Nei Kato
School of Information Sciences
Tohoku University
Sendai, Miyagi, Japan

ISSN 2191-8112 ISSN 2191-8120 (electronic)
SpringerBriefs in Electrical and Computer Engineering
ISBN 978-3-319-25389-3 ISBN 978-3-319-25391-6 (eBook)
DOI 10.1007/978-3-319-25391-6

Library of Congress Control Number: 2015954084

Springer Cham Heidelberg New York Dordrecht London

Printed on acid-free paper

Springer International Publishing AG Switzerland is part of Springer Science+Business Media (www.springer.com)

Preface

The smart grid is still evolving and its potential landscape is enormous. As a consequence, the level of complexity of the smart grid is unprecedented. This book aims to illustrate the highly anticipated opportunities and challenges of the evolving smart grid. It is essential to identify the scope and challenges of the smart grid in a comprehensive manner so as to ensure efficient delivery of sustainable, economic, and secure electricity supplies. The smart grid is the fusion of two core disciplines, the power grid and the communication networks. Therefore, the motivation behind this book is to bridge the gap between the research on power grid systems and on the wireless communication networks through the smart grid concept. In this vein, this book provides an overview of the smart grid and its key advances in architecture, distribution management, demand-side response and load balancing, smart automation, electric storage, power loss minimization, and security. This book further shows that many of the smart grid challenges can be formulated as trade-off problems of the utility operator and its customers. This book also demonstrates how such trade-off problems could be solved using state-of-the-art optimization techniques such as game theory. Numerical results are also included in the book to show the effectiveness of the adopted approaches. We believe that the adopted approaches and the related findings will reveal useful insights for the design of smart grid and spur a new line of thinking for the performance evaluation of evolving smart grid systems.

Acknowledgments

The authors would like to acknowledge the support of the Ministry of Education, Culture, Sports, Science & Technology of Japan, and Tohoku University, Japan. A very special thanks to Prof. Xuemin (Sherman) Shen, the SpringerBriefs Series Editor on Wireless Communications. This book would not be possible without his kind support during the process. Thanks also to the Springer Editors and Staff, all of whom have been extremely helpful throughout the production of this book.

Sendai, Miyagi, Japan   Zubair Md. Fadlullah
                                                                          Nei Kato

Contents

List of Figures

Chapter 1
Introduction

1.1 Evolution of Smart Grids

This book discusses the smart grid [1–8]. The notion smart grid, linguistically, implies an electric power delivery grid having some level of "smartness." Why do we consider these two elements then? First, consider the power system delivery grid. It has frequently been referred to as the greatest and most complex machine ever built in human civilization. In fact, the power grid is almost entirely a mechanical system, with only modest use of sensors, minimal electronic communication, and almost a total lack of electronic control. In the last two to three decades, almost all other industries in the western world have significantly become modernized with the utilization of sensors, communications, and computational ability leading to far greater improvement in efficiency, productivity, service quality, and environmental performance. The power grid system has not evolved, however. There was simply not much of a need for it to change because of the crude way this massive system works with a plethora of components including wires, cables, towers, transformers, circuit breakers, and many more you can possibly think of! When it has needed expansion, more of these components have just been bolted together without paying much attention to what may happen in future if the system needs sophistication or refurbishment. As the power grid started to become bigger in size, its complexity in terms of monitoring and repairing also increased significantly and manual control of the system gradually went out of scope. The industry initiated the use of automated monitoring and power system control with the help of computers began way back in the sixties. This, coupled with a modest use of sensors, has increased over time. Still, however, it is far from being ideal as the power companies lack real-time information on the customer demands and, therefore, are unable to smartly decide power generation and distribution scheduling tasks. The conventional grids, therefore, need to become better and smarter; but investors do not see much profit in

© The Author(s) 2015
Z.Md. Fadlullah, N. Kato, *Evolution of Smart Grids*, SpringerBriefs
in Electrical and Computer Engineering, DOI 10.1007/978-3-319-25391-6_1

1

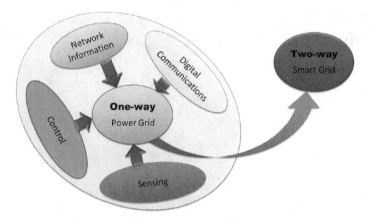

Fig. 1.1 Evolution of the power grid into a smarter "two-way" communication-capable one

retrofitting the old grid with older metering equipment and limited communication capability. This is why there is a lot of stress on the conventional power grid systems to evolve into next generation smart power grid.

So, what contrasting features should the smart grid have? Unlike the traditional grid, the smart grid aims to exploit sensors, communications, computational ability and control in some form to enhance the overall functionality of the electric power delivery system. You may, therefore, define the smart grid as a more naive power grid system evolving into a smarter one through sensing, communicating, applying intelligence, and executing control/feedback for adjusting dynamic changes in primary and renewable electricity supply and demand. The growing trend to converge toward such a two-way communication-capable smart grid is summarized in Fig. 1.1. For a power system, this can effectively allow optimization towards objectives that assure reliability, minimize cost, minimize energy consumption, manage resources, and alleviate environmental impact [9].

The smart grid concept is synonymous to intelligent grid or future grid, and is aimed at providing the end-users[1] with more stable and reliable power. The common aspect of the different smart grid proposals consists in the two-way communication framework between the power source and the consumers, and intelligent sensing entities and control systems lie along the path of the power source and the end-users. The sensors have the ability to detect malfunctions or deviations from normal operational trends which usually require appropriate responses from the smart grid control center. Furthermore, the responses from the control center need to be converted into appropriate control messages and transmitted to different segments of the smart grid. Therefore, the smart grid communication framework and functionality should be characterized, particularly in terms of its capability of

---

[1]The end-users of the smart grid may also be referred to as simply the users, or as the customers or consumers in the book.

provisioning consumer participation, energy demand and cost estimation, demand and load balancing, power loss minimization across micro-grids, and resiliency against malicious threats such as physical and cyber attacks. In a smart grid, consumers are usually equipped with smart meters, which are able to identify power consumption in a much more elaborate way in contrast with the conventional energy meters. While the information gathered by a smart meter can be obtained by the power company (i.e., the utility provider) for monitoring and billing purpose, the consumer can also access his or her smart meter to verify the current use and accordingly adjust the power usage during peak and off-peak hours. In order to do so, the users may have to periodically send request messages to the utility provider. By this way, they can actively and dynamically get involved with the power dissemination process in the smart grid. To implement the two-way communication between the consumers and the utility provider, the smart meters are often integrated in the smart grid by using the so-called AMI, which stands for Advanced Metering Infrastructure. The AMI-based smart meters can be exploited to collect customer demand profiles and forecast their future demand trends so as to project the possible demand ahead of time. Furthermore, it should be stressed that a strong authentication mechanism is a requisite for end-users and smart meters deployed in the smart grid. Such an authentication mechanism needs to be fast yet secure so that the smart grid entities may be able to quickly verify each other's identities and to allow the users to access smart grid services and resources in a secure manner.

With these multi-directional challenges, even though the smart grid has been gaining momentum in recent years, researchers are still trying to imagine and predict how an actual full-scale smart grid communication framework and structure should be. As a matter of fact, the smart grid vision can only be realized through the fusion of various disciplines where professionals from power engineering background will meet the communication engineers. To determine which technologies from these various disciplines should be exploited to get the best return has been a daunting task. For example, do we need to enhance the currently available communication technologies or grasp the upcoming 5G technologies to suit the smart grid requirements? Can and how the communication technologies help the smart grid to reduce power loss? How can the smart grid be secure against well-known easily doable spoofing and man-in-the-middle attacks? Even though this book cannot give answers to all the challenges and issues related to the smart grid, we would like to cover some of the most basic yet challenging problems and address those specific problems to gain some steps towards enabling a reliable and secure smart grid.

1.2 Challenges with Current Smart Grid Settings

Even though there appeared quite a few researches in the smart grid literature, the available approaches to address the smart grid communications are still somewhat limited. This means that a lot of work needs to be re-addressed to provide a reasonable smart grid models. In other words, the first and foremost agenda is to agree (or not to agree) on what should be the most realistic power distribution and communication model for a practical and scalable smart grid. Why is this important? Because, a smart grid which will cost billions of dollars needs to be designed in a very careful manner from the beginning. The smart grid will have numerous electrical appliances connected with each another in a uber-complex fashion so that they can report back on elements like power consumption and other monitoring signals. So, fast communication is crucial yet challenging for practical deployment of the smart grid. In order to realize effective smart grid communication, existing networking technologies need to be exploited (and even enhanced) to cope with the multiple services and quality requirements of the residential appliances. The need to distinguish between the high- and low-priority traffic will be equally important as to be able to dynamically adapt the network to changing capacity demands in real-time. Therefore, it is essential that we consider appropriate technologies to implement the communication networks of smart grid, which may allow the flexible use of existing capacities without impacting the service quality of the smart grid.

The renewable electricity generation and delivery is another important aspect of the smart grid. Micro grids (smaller autonomous power generation and dissemination systems) which could be connected and isolated from the main power grid are becoming more common. The micro grids are perceived as the means to reduce unnecessary/fossil-based power generation and distribution, and reduce the total power loss across the main grid. The echo-friendly or "green" micro grids are, therefore, gaining lots of attention from the mainstream smart grid community. But when you deal with a large number of micro grids, substantial reduction in power loss may be impossible without optimizing the scheduling of local energy generation and distribution of the micro grids. In addition, the micro grids could have local storage facilities to locally store energy for selling to their own customers or even other micro grids at a later time. Should we consider the power generation and storage losses also in addition to the distribution loss? In addition, how to deal with such problems? In other words, who will decide the schedule of generation and storage? A centralized decision maker certainly sounds interesting, but may not be as impressive in real-time compared to a distributed decision making solution.

Security is another key challenge of the smart grid. The Internet Protocol or IP-based communication adopted in the smart grid communication means the malicious threats plaguing the Internet could be very much possible in the smart grid. The communication links between the smart meters and the control centers, therefore, need to be secure against various security threats. Given the volume of messages, delay constraints, and limited hardware capability of the smart meters, a light-weight message authentication scheme tailored for the smart grid could be greatly useful.

1.3 Aim of the Book

The book mainly focuses on the anticipated opportunities and challenges of the smart grid. In Chap. 2, we will provide our definition of a smart grid system model from two perspectives. The first perspective considers an overlay of the power distribution network and the communication network-based hierarchical smart grid framework. The second perspective describes a micro grids-based architecture with a focus on micro grids equipped with renewable energy sources and storage devices. We will investigate and compare the performance of these two models and discuss whether a fusion of the two could be possible.

In Chap. 3, the challenges of the smart grid will be more deeply visited. Most of these challenges can be considered as optimization problems. So, in the chapter, we will describe how existing works formulated these optimization problems from different perspectives.

From Chap. 4 onward, the book will transform from a broader narrative to a more technical discussion. In Chap. 4, we will describe a demand-side management based on game-theory and its advantages in improving the peak-to-power ratio. In Chaps. 5–7, we will discuss power loss reduction strategies in micro grids. These contents of these three chapters are incremental. Chapter 5 will describe the power loss model of a basic (theoretical) micro grids based system and show how a centralized game-theoretic approach can effectively improve the power loss. In Chap. 6, the micro grid model is extended to a more practical one comprising power storage devices. The chapter shows with this slight extension, the problem becomes even deeper which needs even more sophisticated algorithm to have a central decision making. In Chap. 7, we move away from the central decision making strategy to a distributed paradigm for power loss reduction in micro grids so as to increase resiliency of the individual micro grids.

In Chap. 8, we shift to a different yet relevant topic where we discuss the need of a secure framework for smart grid communications. The chapter will introduce a secure and reliable framework for smart grids, a key part of which is a lightweight and secure message authentication scheme tailored for the requirements of smart grid's AMI. The chapter adopts a Diffie–Hellman key establishment protocol and hash-based message authentication code, which lets various smart meters at different segments of the smart grid to establish mutual entity authentication with low delay and communication overhead.

For interested readers, in the technical chapters (i.e., Chaps. 4–8), technical analysis is shown to indicate how the adopted approaches can satisfy the desirable requirements of the respective problems. Some performance evaluation is also included in the chapters to demonstrate the effectiveness of the adopted approaches.

Finally, the book is concluded in Chap. 9 by summarizing the various challenges which are critical for a smart grid technology to evolve and mature. Future directions and caveats are also described in the section.

References

1. M. M. Fouda, Z. M. Fadlullah, N. Kato, R. Lu, and X. Shen, "A lightweight message authentication scheme for smart grid communications," *IEEE Transactions on Smart Grid*, vol. 2, no. 4, pp. 675–685, Dec. 2011.
2. H. Liang, A. Tamang, W. Zhuang, and X. Shen, "Stochastic information management in smart grid," *IEEE Communications Surveys Tutorials*, vol. 16, no. 3, pp. 1746–1770, Mar. 2014.
3. M. M. Fouda, Z. M. Fadlullah, and N. Kato, "Assessing attack threat against zigbee-based home area network for smart grid communications," in *2010 International Conference on Computer Engineering and Systems (ICCES)*, Cairo, Egypt, Nov.-Dec. 2010, pp. 245–250.
4. H. Liang, B. J. Choi, W. Zhuang, and X. Shen, "Stability enhancement of decentralized inverter control through wireless communications in microgrids," *IEEE Transactions on Smart Grid*, vol. 4, no. 1, pp. 321–331, Mar. 2013.
5. M. M. Fouda, Z. M. Fadlullah, N. Kato, R. Lu, and X. Shen, "A lightweight message authentication scheme for smart grid communications," *Smart Grid, IEEE Transactions on*, vol. 2, no. 4, pp. 675–685, Dec. 2011.
6. Z. M. Fadlullah, M. M. Fouda, N. Kato, A. Takeuchi, N. Iwasaki, and Y. Nozaki, "Toward intelligent machine-to-machine communications in smart grid," *IEEE Communications Magazine*, vol. 49, no. 4, pp. 60–65, Apr. 2011.
7. A. Abdallah and X. Shen, "Lightweight security and privacy preserving scheme for smart grid customer-side networks," *IEEE Transactions on Smart Grid*, to appear.
8. H. Li, X. Lin, H. Yang, X. Liang, R. Lu, and X. Shen, "EPPDR: an efficient privacy-preserving demand response scheme with adaptive key evolution in smart grid," *IEEE Transactions on Parallel and Distributed Systems*, vol. 25, no. 8, pp. 2053–2064, Aug. 2014.
9. C. W. Gellings, *The smart grid: Enabling energy efficiency and demand response*. Fairmont Press, 2009.

Chapter 2
Considered Smart Grid Model

Before discussing the challenges of the smart grid (and even addressing them), to have an understanding of how the smart grid is modeled and how it operates would be useful [1–4]. That is basically the aim of this chapter. To make it interesting to the readers, two slightly different approaches to the smart grid model are covered in the chapter. The first one is more like an overlay between the power grid and the communication system [1]. Hierarchical communication networks are used in this first approach to allow the bi-directional flow of information between the utility provider and the customers. Smart meters and AMI are the core technologies adopted in this case. In the second approach, a micro-grids based smart grid architecture [5] is considered where the communication networks are somewhat inline which assume the use of either dedicated or already existing communication technologies or even the Internet.

2.1 Smart Grid Model with Hierarchical Communication Networks

Figure 2.1 demonstrates our considered smart grid communication framework which has been adopted in [1, 2, 6]. The smart grid power transmission and distribution system is considered in the figure to be separated from the communication system. First, let us discuss the power distribution network briefly. The power generated at the power plant(s) is supplied to the customers via the transmission and distribution substations. The transmission substation is located usually near the power plant. On the other hand, the distribution substations are located in different areas and neighborhoods. The transmission substation delivers electric power over high voltage (>230 kV) transmission lines to the distribution substations

© The Author(s) 2015
Z.Md. Fadlullah, N. Kato, *Evolution of Smart Grids*, SpringerBriefs
in Electrical and Computer Engineering, DOI 10.1007/978-3-319-25391-6_2

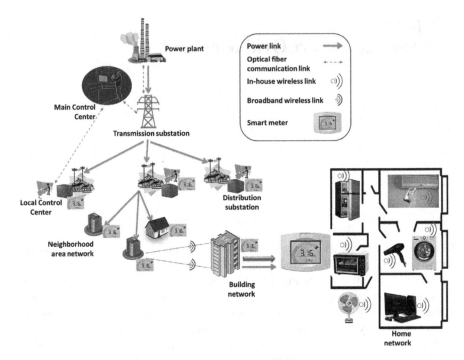

Fig. 2.1 Considered smart grid model with hierarchical communication networks

that transform the power into medium voltage level and then distribute it to the building-feeders. The building-feeders convert the medium level voltage into a much lower one, which becomes usable by user appliances.

From the communication perspective, the smart grid architecture is divided into a number of hierarchical networks. The transmission substation near the power plant, and the control centers of the distribution substations are connected with each another. The mesh network that connects them is typically implemented using high speed optical fiber to support high volume of smart grid data traffic with the least possible communication delay.

On the other side of the control centers, the communication framework is divided into a number of hierarchical networks such as neighborhood area networks (NANs), building area networks (BANs), and home area networks (HANs). For simple understanding, consider that each distribution substation covers a single neighborhood where a NAN is established. Each NAN has a number of BANs. Each BAN has multiple HANs (basically several apartment-based networks). The smart meters are the key devices in this communication architecture that act as the dual-gateway (GW) for energy use monitoring and communication. Depending on where they are placed, the smart meters are referred to as the NAN GWs, BAN GWs, and HAN GWs, and they are not necessarily of the same specs and hardware. For example, a BAN GW, installed alongside a building feeder, can have more sophisticated hardware and specs, compared to a HAN GW. On the other hand, through a cheaper and lower-spec HAN GW, the customer can check energy use and express his/her energy demand to the grid operator at the control center. A smart meter having the MSP430F471xx micro-controller can be used as a typical HAN GW [7] with a random access memory of just 8 kilo bytes and flash memory of 120 kilo bytes. The HAN GW has a 16 MHz CPU, 3/6/7 16-bit analog to digital converters (ADCs) and programmable gain amplifiers (PGAs), 160-segment liquid crystal display (LCD), real-time clock (RTC), and 32×32 hardware multiplier. On the other hand, much more powerful (up to ten times more capability) smart meters are needed to act as BAN GW with specs like 160 MHz CPU, 128 kilo bytes random access memory, and 1 mega byte flash memory. As for the NAN GW configuration, a dedicated computer with Intel Core haswell-i7 CPU, 16 giga bytes random access memory, and storage scaling up to few terabytes could be considered. The difference in these smart metering specifications is because the end-users are likely to deal with significantly lower traffic and usually prefer affordable (i.e., low cost) smart meters whereas the grid operator at the control center can easily afford one or more high-spec computers to cope with substantially large volume of data generated from a large number of customers in the neighborhood.

Based on the existing standards of smart grid, IP-based communication is preferred to allow seamless inter-connections with HANs, BANs, NANs, control centers, and the transmission substation. What kind of IP-communication technology should be used at each hierarchical network of the smart grid is described below.

2.1.1   Communication Technology Used in HAN

In the considered smart grid, HAN is basically the "last mile." The HAN allows the end-users to easily manage their energy use and demand. Let us refer to the HAN shown on the bottom right of Fig. 2.1. The home network connects the smart appliances (e.g., television, washing machine, oven, etc. which have their own IP address) to the HAN GW, which can consistently or periodically communicate with *BAN*1. Smart Energy Profile (SEP) Version 1.5 over IEEE 802.15.4 ZigBee radio communications could be a candidate HAN communication technology because of its low power requirements (1–100 mW) compared to other home-equipment radio technologies like the IEEE 802.11 (WiFi) and Bluetooth. Also, ZigBee provides a reasonable communication range of 10–100 m and incurs moderate cost, and therefore, it can be considered as the HAN-enabling communication technology. However, placement of the HAN GW inside the apartment to establish stable connectivity with all the equipment is still critical.

2.1.2   Communication Technology Used in BAN

Just like a building has several apartments, a BAN has a number of HANs. The BAN GW at the building feeder can be used to monitor the energy consumption and project the demand of the building's residents. Although WiFi could appear to be a common choice due to its common availability in recent times, a BAN covering many households in a large apartment building means potential lack of WiFi coverage. Deploying multiple WiFi access points could be a solution which may still be subject to serious interference from other access points commonly used for Internet access and so forth. Therefore, 4G technologies like LTE or WiMax could be harnessed to cover more areas to facilitate communication between a BAN and its covered HANs.

2.1.3   Communication Technology Used in NAN

A NAN connects the smart grid's backhaul with the radio access networks (i.e., BANs and HANs). The NAN signifies a specific area like a ward of a city. Through the NAN GW, the grid operator monitors how much power is being distributed to a particular neighborhood by the respective distribution substation. Broadband wireless technologies such as 4G LTE and WiMax can be exploited for connectivity of the NAN GW and its covered BAN GWs. It is worth noting, however, that the existing studies indicated smart grid communication network to be separate from the existing ones for providing Internet services so as to prevent network congestion and malicious threats.

2.2   Smart Grid Communication Packet Structure

The smart grid data are exchanged between the customers' smart meters and the control center through the HAN, BAN, and NAN GWs using IP-based communication. So, it is important to consider the smart grid data communication packet structure from a high level as shown in Fig. 2.2. The illustration in the figure is based on commercial smart meter specifications in [8]. In addition to the raw message, every packet has the message header, TCP/IP header, and security header. The message header includes meter ID MAC address, equipment status, and the Type of Message (ToM). There are nine possible ToMs, which the HAN GW is able to send to the BAN GW. The function and size of each ToM are also indicated in the figure.

A nuclear power plant could also be an example of the macro station.

Sample structure of a communication packet
in the considered smart grid

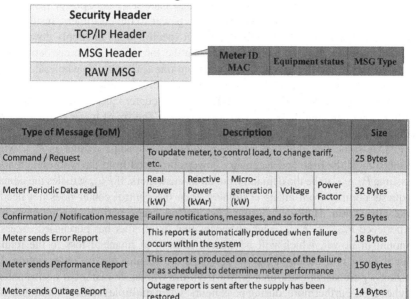

	Type of Message (ToM)	Description					Size
1	Command / Request	To update meter, to control load, to change tariff, etc.					25 Bytes
2	Meter Periodic Data read	Real Power (kW)	Reactive Power (kVAr)	Micro-generation (kW)	Voltage	Power Factor	32 Bytes
3	Confirmation / Notification message	Failure notifications, messages, and so forth.					25 Bytes
4	Meter sends Error Report	This report is automatically produced when failure occurs within the system					18 Bytes
5	Meter sends Performance Report	This report is produced on occurrence of the failure or as scheduled to determine meter performance					150 Bytes
6	Meter sends Outage Report	Outage report is sent after the supply has been restored					14 Bytes
7	Weekly read submission	Output data after one week					28 Bytes
8	One month of data	Meter sends one month of data					40320 Bytes
9	Last day import data	Summary of usage on the last day					192 Bytes

Fig. 2.2 SG communications packet structure

2.3 Micro Grids Oriented Smart Grid Model

This section of the chapter will describe a different perspective of the smart grid architecture based on micro grids and its collaboration with the macro station. Let us first discuss the macro station and its relationship with the retailers before grasping the concept of micro grids [5, 9–11].

2.3.1 Macro Station and Retailers

The smart grid is a cyber-physical system as it combines the power grid and communication system (you should understand this by now if you have carefully followed Sect. 2.1). The detailed information of the smart grid communication architecture was given in Sect. 2.1, and here, we slightly tilt our perspective to the power side of the smart grid. The wholesale power market is composed of the main power plant (i.e., geo-thermal/hydro-electric/nuclear power plants) and the renewable power generation sources (e.g., solar panels, wind turbines, hybrid car batteries, etc.), which produce the primary (base) and secondary power, respectively. The wholesaler does not directly supply the power to the customers. Instead, there are power "retailers" that usually operate in specific areas to supply power to the customers living in those areas. Based upon the forecast of customer demands for the next day, a retailer purchases power from one (or multiple) wholesaler(s), and sells this power to its own customers. Every retailer has a control center, which you could possibly relate with Sect. 2.1. The control center carries out authentication, billing, data aggregation, and direct load control. The smart meter of the customer and the control center can communicate over wireless technology (Sects. 2.1.1–2.1.3). Assume that the total power available at a specific retailer is P_{max} which is the sum of the base and secondary power purchased from the wholesale market. Assume P_{demand} to be the demand from all its customers. If P_{max} satisfies the demand requests coming from all its customers, then the control center does not need to take any action. However, during low-demand usage periods, the retailer could experience $P_{max} \gg P_{demand}$ which means that the retailer has purchased excessive amount of power, which will be wasted leading to financial loss to the retailer. But if there is a deficit ($P_{max} < P_{demand}$), the retailer has no other option than instantly purchasing additional power from the wholesale market at a substantially higher cost. Furthermore, during peak hours, when multiple retailers simultaneously want to buy additional power, the primary source may have already exhausted its capacity. Then, the retailers would be forced to buy the extra power from, perhaps other retailers, which may be even more expensive. Intuitively, in such a scenario, a retailer's best option would be to seek secondary sources in its locality and purchase power from them. These localized secondary sources, which some refer to as the micro grids, are recently becoming popular.

2.3.2   Macro Station and Micro Grid Model

In this model, the macro station (some refer to it as the MS) represents the wholesaler of the energy market which usually generates electricity from fossil-based, non-renewable resources like coal, petroleum, etc. Such a macro station deals with a high maintenance cost due to somewhat inefficient fuel to electricity generation and high carbon footprint. Yes, you could possibly imagine a nuclear power plant as a macro station; but after the 2011-tragedy at Fukushima where the Tokyo Electric Power Company (TEPCO) had its nuclear power-driven macro station for the greater north-east Japan, interest toward spatially distributed, smaller yet independent micro-grids capable of generating and serving their own power from renewable sources has become paramount. In addition to dealing with emergency demand in post-disaster scenarios, the micro grids are getting a great deal of interest from the US military. Compared with the conventional power grids, the smart micro grids are more flexible since they can quickly adjust their power production in

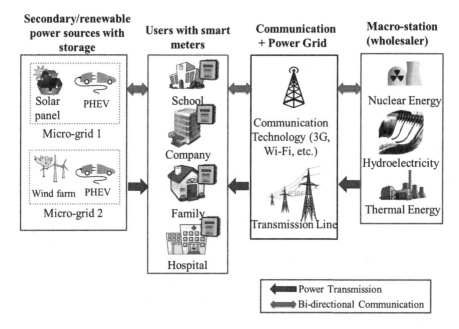

Fig. 2.3 Micro-grids oriented architecture of the smart grid

accordance with user-demand. When a micro-grid is not self-sufficient, it can buy extra power from the wholesaler or even the neighboring micro grids for satisfying the energy demands of its customers. Thus, the micro grids can act as a viable complement to the conventional macro station. The micro-grids based smart grid framework is broadly shown in Fig. 2.3. Readers can check an earlier work of micro grids based approach for the power distribution system with optimized self-adequacy in [12].

2.3.3 Distributed Micro Grids with Renewable Sources

The end-users of a micro grid may be typical consumers like residential users, companies, schools, hospitals, government facilities, and so forth. The end-users can get power from their subscribed micro grid(s). Because most micro grids these days have integrated renewable resources like wind, solar, water turbine, etc., in addition to the purchased power from the macro station. Micro grids exploiting renewable sources may not always be able to have a stable production and supply of electricity. This happens because of the complex variation of power demand of the users depending on the different times of a day [13, 14], the power demand of the users varies depending on the different times of a day. Particularly during the peak hours when the electricity consumption of the users is the highest, the micro grids have to ensure a stable power supply even though instantaneous production is not possible. So, how can a micro grid cope with the total user-demand approaching or even exceeding its supply power? The micro grid has two options. First, the micro-grid can buy additional power from the macro station and/or from other micro grids. Second, the micro grid can have dedicated power storage devices like batteries, plug-in hybrid electric vehicles (PHEVs), etc., which may be charged during the off-peak hours and discharged during the peak hours to meet the user-demand. These two techniques are not necessarily mutually exclusive.

2.4   Deployment Scenarios and Key Opportunities

The review of available micro grid architectures showed that most of the existing testbeds are AC (alternating current) micro grids [15] because they are easy to integrate with the main power grid and most loads (user appliances) which run on AC. The AC micro grids, however, struggle with maintaining a stable power quality. This is one of the critical shortcomings of the AC micro grids. In contrast, the DC (direct current) systems in general are not prone to power quality problems. But deploying the DC micro grid is not practiced much due to limited use of DC loads by the customers. HFAC (high frequency AC) micro grids are now gaining popularity that can help integrating renewable energy sources with the micro grid while maintaining a reasonable power quality. A HFAC micro grid system has its own disadvantage, however. It usually needs more control devices, and suffers from a large voltage drop and long distance power loss. The conventional power generation source of the micro grids is typically diesel; but renewable sources like solar power volt, wind, and micro-hydro systems are becoming popular deployment choice for the micro grids. Because the renewable sources are highly dependent on the ambient environment, the micro grid usually is deployed with power storage devices to store power for later use. Most existing test-beds have battery storage while some have capacitor banks and flywheels as storage devices. Several storage units are deployed in some micro grids while others do not have any storage unit at all. If the micro grid system does not have any storage device and only have renewable energy source, then the main grid connectivity is a very important option for that micro grid system. There is definitely room for further study on the deployment scenarios of the micro grids.

To demonstrate that micro grids are not merely research projects, we include an example of a major aversion of power outage by exploiting a micro grid using renewable sources in late May 2015 [16]. The Borrego Springs micro grid supplied energy to 2800 customers for 9 h until the utility operator (San Diego Gas & Electric) repaired the damage inflicted by a lightning strike. The micro grid automatically switched between its different power generation sources like onsite energy, energy storage, and a 26 MW solar generation source to steadily manage power supply to the customers of the entire community. This shows the immense potential of micro grids in days to come.

2.5 Practical Implementation and Operating Challenges

The DC micro grids are not that popular in European countries even though it does not exhibit much power quality issues. With the advancement of technology, perhaps more DC-compatible loads and appliances can be designed and implemented to encourage using the DC micro grids. In case of the HFAC micro grids, the main barrier to practical implementation is the significantly expensive solar cells and storage devices. The operating challenge is also to be taken into account since the clean power generation from renewable sources depend heavily on the environment. Currently, the initial implementation cost along with operation and maintenance cost are on the higher side. So, more penetration of renewable sources is expected in the micro grid systems and further technological improvement is needed to make them an economically viable option.

At the same time, we have to keep an open eye for the telecommunications needs for distributed generation in the overall smart grid. The best policy in distribution planning is to anticipate communications requirements and plan in a comprehensive way [17]. With the advent of 5G communication technologies by 2020, many operators would want to take advantage of those and meanwhile just use what is available to them—LTE, LTE-A, or WiMax broadband technologies in the radio access part (HAN–BAN and BAN–NAN communication). Therefore, there needs to be a very comprehensive strategy so as to make a seamless transition to the newer technologies. Otherwise, there could be adverse effects such as installation of incompatible or redundant systems.

References

1. M. M. Fouda, Z. M. Fadlullah, N. Kato, R. Lu, and X. Shen, "A lightweight message authentication scheme for smart grid communications," *IEEE Transactions on Smart Grid*, vol. 2, no. 4, pp. 675–685, Dec. 2011.
2. M. M. Fouda, Z. M. Fadlullah, N. Kato, R. Lu, , and X. Shen, "Towards a light-weight message authentication mechanism tailored for smart grid communications," in *IEEE International Workshop on Security in Computers, Networking and Communications (SCNC'11)*, Shanghai, China, Apr. 2011.
3. H. Liang, A. Tamang, W. Zhuang, and X. Shen, "Stochastic information management in smart grid," *IEEE Communications Surveys Tutorials*, vol. 16, no. 3, pp. 1746–1770, Mar. 2014.
4. H. Liang, B. J. Choi, W. Zhuang, and X. Shen, "Stability enhancement of decentralized inverter control through wireless communications in microgrids," *IEEE Transactions on Smart Grid*, vol. 4, no. 1, pp. 321–331, Mar. 2013.
5. C. Wei, Z. M. Fadlullah, N. Kato, and A. Takeuchi, "GT-CFS: a game theoretic coalition formulation strategy for reducing power loss in micro grids," *IEEE Transactions on Parallel and Distributed Systems*, vol. 25, no. 9, pp. 2307–2317, Sep. 2014.
6. Z. M. Fadlullah, M. M. Fouda, N. Kato, A. Takeuchi, N. Iwasaki, and Y. Nozaki, "Toward intelligent machine-to-machine communications in smart grid," *IEEE Communications Magazine*, vol. 49, no. 4, pp. 60–65, Apr. 2011.
7. Texas instruments - msp430 for utility metering applications. Available online, accessed Aug 2015. "http://focus.ti.com/mcu/docs/mcuorphan.tsp?contentId=31498".

8. "High-level smart meter data traffic analysis," Engage Consulting Ltd. for the Energy Networks Association (ENA), UK, Tech. Rep., May 2010.

9. P. Aristidou, A. Dimeas, and N. Hatziargyriou, " Microgrid modelling and analysis using game theory methods," in *Energy-Efficient Computing and Networking*, ser. Lecture Notes of the Institute for Computer Sciences, Social Informatics and Telecommunications Engineering, N. Hatziargyriou, A. Dimeas, T. Tomtsi, and A. Weidlich, Eds. Springer Berlin Heidelberg, 2011, vol. 54, pp. 12–19. [Online]. Available: http://dx.doi.org/10.1007/978-3-642-19322-4_2

10. C. Wei, Z. M. Fadlullah, N. Kato, and I. Stojmenovic, "On optimally reducing power loss in micro-grids with power storage devices," *IEEE Journal on Selected Areas in Communications*, vol. 32, no. 7, pp. 1361–1370, Jul. 2014.

11. ——, "A novel distributed algorithm for power loss minimizing in smart grid," in *2014 IEEE International Conference on Smart Grid Communications (SmartGridComm)*, Venice, Italy, Nov. 2014, pp. 290–295.

12. S. Arefifar, Y. Mohamed, and T. El-Fouly, "Supply-adequacy-based optimal construction of microgrids in smart distribution systems," *IEEE Transactions on Smart Grid*, vol. 3, no. 3, pp. 1491–1502, Sep. 2012.

13. X. Liang, X. Li, R. Lu, X. Lin, and X. Shen, "UDP: usage-based dynamic pricing with privacy preservation for smart grid," *IEEE Transactions on Smart Grid*, vol. 4, no. 1, pp. 141–150, Mar. 2013.

14. S. Chen, N. Shroff, and P. Sinha, "Heterogeneous delay tolerant task scheduling and energy management in the smart grid with renewable energy," *IEEE Journal on Selected Areas in Communications*, vol. 31, no. 7, pp. 1258–1267, Jul. 2013.

15. L. Mariam, M. Basu, and M. F. Conlon, "A review of existing microgrid architectures," *Journal of Engineering*, vol. 2013, no. Article ID 937614, pp. 1–8, 2013.

16. E. Wood, "Borrego Springs Microgrid Averts Long Outage During Lightning Strike Repair," (accessed Aug. 2015). [Online]. Available: http://microgridknowledge.com/borrego-springs-microgrid-averts-long-outage-during-lightning-strike-repair/

17. J. Brandt, "Is telecom a natural transition for utilities?" accessed Aug. 2015. [Online]. Available: http://www.smartgridnews.com/tech/story/telecom-natural-transition-utilities/2015-08-11

Chapter 3
Challenges in Smart Grid

Several smart grid pilot projects have been developed in recent year. The rapidly evolving scenario of smart grids with distributed generation and storage means dealing with multi-directional challenges involving higher communication capabilities, faster reporting rates, better security, lower power losses, and so forth [1]. Understanding these challenges and then adequately addressing them are a prerequisite for the complex control and management applications of the smart grid. In this chapter, we describe some of the critical optimization challenges of the smart grid from different perspectives.

3.1 Optimization Challenges from Different Perspectives

3.1.1 Demand Response (DR): Shaping Electricity Demand to Match Supply

The smart grid operators need to guarantee that there is sufficient power generation to service the load in terms of both wattage and volt-amperes reactive. The challenge can be simply summarized as the classical supply and demand problem. How can the smart grid keep supply and demand in balance? The traditional way to achieve this is to continually adjust central generation through demand-side management (DSM) [2–5], which includes energy efficiency and demand response (DR). Instead of simply adding more generation to the system, utility operators pay the end-users to reduce energy consumption since this is easier and cheaper than traditional generation. To accomplish this, an operator can offer DR programs to encourage its customers to make short-term reductions in energy demand as a response to a price signal from the operator to switches on customers' loads (like air-conditioners, washing machines, water heaters, lighting arrays, etc.). Usually the DR actions would be in the range of a few hours.

© The Author(s) 2015
Z.Md. Fadlullah, N. Kato, *Evolution of Smart Grids*, SpringerBriefs
in Electrical and Computer Engineering, DOI 10.1007/978-3-319-25391-6_3

3.1.1.1 Direct Load Control

The direct load control (DLC) is an example of DR programs which have helped throttle back peak demand. DLC examples include shutting down loads, cutting off industrial process, or cycling it off for a specific time. By this simple yet effective way, the operator may survive a few critical peak hours instead of having to provide additional generation. Traditional DLC methods depend on legacy communications, however, that lack accurate forecast and customer participation which are necessary to optimize the DR programs. Also, some customers have privacy concerns regarding DLC simply because they often do not want to give such control to the utility operator (or anyone else for that matter). To overcome this privacy issue, additional policies need to be integrated which could differentiate various types of appliances with different grades of energy usage preferences. A transparent policy can offer the users a clear choice between utilizing or not utilizing DLC. In other words, if a user wants to cut energy bill, he/she can give the utility operator adequate permission to apply DLC.

3.1.1.2 Price-Based DR Programs

Agarwal and Cui [6] argue that charging the customers a flat rate for energy use leads to "allocative inefficiencies." To overcome this, smart grid operators can leverage pricing incentives (such as peak time rebates, time-of-use rates and price response) to encourage customers to shift usage to off-peak hours. These price-based DR programs give the customers a clear and up-to-date snapshot of their energy usage and cost, and deliver more control over their energy bill. Theoretically such programs sound great because they assume that all the customers will cooperate. But in reality whether the users will accept the incentive or not is also dictated by daily habits, psychology, awareness, and other social factors.

3.1.2 Energy Conservation and Efficiency

Energy conservation (load reduction) and efficiency are other DSM techniques which the smart grid operators can exploit. Energy conservation programs are used to encourage the users to sacrifice some energy use. In other words, they reduce the overall demand for electricity by reducing the amount of utility the customer receives. Example of this includes turning up the thermostat a few degrees in summer in order to reduce air-conditioning. On the other hand, the energy efficiency programs aim to reduce the overall demand for electricity while maintaining the amount of utility the customer receives. For instance, when a user replaces him/her old air-conditioner with a more energy-efficient model.

How energy demand and supply estimation can be more effectively balanced has, thus, become a topic of interest. An autonomous and distributed demand-side energy management system using game theory is an early work [7] in this area.

This work shows how game theory has recently emerged as a promising technique to solve such trade-off of power supply and demand so that both the utility operator and customers may benefit. These game-theoretic approaches allow the customers negotiate with the power company or operators until they reach an equilibrium, a point of "consensus" so to speak, where all the stakeholders feel they have made the best of the negotiation. The problem is depending on how the optimization problem or the game is formulated, such a consensus may take a long time to reach, or it may not at all exist in some cases.

Furthermore, the DSM techniques, particularly the DR programs, usually only consider either the interest of the utility company or the customer. Consider a simple example. A power company wants to maximize its revenue while its customers want to reduce their energy costs. If so, for example, a program that prices peak electricity up and discourages consumption might actually deprive the utility of revenue. This case is just one of many in which contrasting demands make it difficult to design a joint DSM technique that takes into account the interests of the utility company and its customers simultaneously.

3.1.3 Power Loss Problem in Micro-Grids Based Smart Grid System

There are different types of power losses associated with power exchange in the smart grid. For instance, when power is transmitted, heated power distribution lines result in transmission power loss. Power storage results in storage power loss. Furthermore, different choices might cause different power losses. For example, under the same physical conditions, transmission power loss caused by obtaining power from a nearby micro grid is lower than that from a distant micro grid. Therefore, an efficient algorithm that is able to optimally reduce the total power losses in smart micro-grid system. Note that whether we consider power transmission in the traditional power grid or in the smart micro-grid based power delivery system, power loss is an unavoidable reality.

Even in the state-of-the art HFAC micro grids, large voltage drops and higher long distance power loss are the key issues that currently limit their practical implementation. The power loss figure may not appear to be substantial for just one or two individual micro grids. But when you consider the entire power grid system, the power loss is just too much. [8] reported the power loss of the 20,279,640 gWh power consumed by the whole world in 2009, and it indicated that if there was some way to reduce the power loss even by 1 % of the total power consumption, more than billions of dollars could have been saved.

Recently, there has been a significant stress on studying the issue of power loss between the micro-grids and that between an individual micro-grid and the macro-station recently. In [9], energy management in smart grid powered electrical appliances investigated with a particular emphasis on energy savings from their proposed strategy. Arefifar et al. presented systematic and optimized approaches for the power distribution system into a set of micro-grids with optimized self-adequacy [10]. In their work in [11], Niyato et al. proposed an algorithm which optimizes the transmission strategy to minimize the total cost. The problem of minimizing power losses in distribution networks has traditionally been investigated by using a single, deterministic demand level. Ochoa et al. presented a novel algorithm to solve this problem [12]. Kantarci et al. proposed the "cost-aware smart micro grid network design," which enables economic power transactions within the smart grid [13]. The work in [14] proposed operation strategies to be implemented into the distribution management system for power loss reduction. Some of the other notable research works, which have addressed power loss can be found in [15, 16]. In one of our earlier works [17], a game-theoretic coalition formulation strategy for reducing power loss in micro-grids dubbed as GT-CFS was proposed.

3.2   Is Smart Grid Communication Security an Overkill?

In the smart grid, the customers are no longer passive stakeholders. They can actively get involved to effectively minimize energy consumption by communicating back and forth with the utility operator. Numerous sensing devices, smart meters, and control units are likely to be between the operator and end-users to realize the bi-directional communication. IP-based communication technologies are considered to be the top candidate for setting up HANs, BANs, and NANs. This means the smart meters and loads are all hooked up to the IP-based communication network. However, existing IP-based communication networks (e.g., the Internet) are likely to be susceptible to a wide variety of malicious attacks, such as replay, traffic analysis, and denial of service (DoS) attacks. The smart grid communication network is also no exception. Therefore, it is important to properly design smart grid communication protocols for dealing with all possible security threats [18–20]. However, designing a security protocol from scratch specifically for smart grid communication may not be practical. Using heavy duty encryption mechanisms can actually be an overkill since they can significantly add delay and packet overhead affecting the delay-sensitive communication requirement of the smart grid. The smart grid needs a light-weight secure protocol suite customized to meet its communication needs. As a first line of defense against malicious threats to smart grid communication, a light-weight authentication scheme is required. Just like the traditional IP-based communication networks, the smart grid communication framework needs to check whether the parties involved in communication are the exact entities they appear to be. Hence, the smart grid communication framework should consider an adequate authentication mechanism [21–28] so that malicious attackers cannot compromise the secrecy or privacy of the information exchanged between the utility operator and its customers.

3.3 Concluding Remarks: Could an All-in-One Optimization Technique Work in Smart Grid?

How acceptable and widespread the smart grid system will become depends a lot on how its design can cope with some of the most critical challenges such as balancing electricity demand and supply, and power loss minimization. Careful design to combat these challenges is the key. If the utility company directly controls the residential loads, the customers may perceive this as breach of privacy. If each customer needs to sign a contract with the utility for DLC, this might not get beyond large-scale users like malls, warehouses, and factories. These are just some simple examples to highlight the fact that smart grid challenges are both technical and social, and therefore, they need cautious design and evaluation. Furthermore, note that the smart grids are not blackout-proof. It will work fine under normal operations flattening the load curve, but blackouts can still happen in a smart grid when the power plants fail to operate during the peak demand hours. Hence, even in a smart grid, baseload generation, dispatchable generation through DLC, and storage are required. The smart grid just pushes out the eventuality of a blackout far enough so that most of the grid system can safely recover from intermittent sources without increasing the frequency of blackouts. The concluding remark is that an all-in-one optimization technique in smart grid may not be possible because DSM, power loss minimization, and security have different and sometimes contrasting objectives. Instead of an all-in-one solution, an extensive framework with flexible action-policies can be offered from which the utility operator and customers can select the best course of action.

References

1. H. Liang, A. Tamang, W. Zhuang, and X. Shen, "Stochastic information management in smart grid," *IEEE Communications Surveys Tutorials*, vol. 16, no. 3, pp. 1746–1770, Mar. 2014.
2. M. M. Fouda, Z. M. Fadlullah, N. Kato, A. Takeuchi, and Y. Nozaki, "A novel demand control policy for improving quality of power usage in smart grid," in *2012 IEEE Global Communications Conference (GLOBECOM)*, Anaheim, California, USA, Dec. 2012, pp. 5154–5159.
3. C. Gellings, "The concept of demand-side management for electric utilities," *Proceedings of the IEEE*, vol. 73, no. 10, pp. 1468–1470, Oct. 1985.
4. Z. M. Fadlullah, M. Q. Duong, N. Kato, and I. Stojmenovic, "A novel game-based demand side management scheme for smart grid," in *2013 IEEE Wireless Communications and Networking Conference (WCNC)*, Shanghai, China, Apr. 2013, pp. 4677–4682.
5. Z. M. Fadlullah, D. M. Quan, N. Kato, and I. Stojmenovic, "GTES: an optimized game-theoretic demand-side management scheme for smart grid," *IEEE Systems Journal*, vol. 8, no. 2, pp. 588–597, Jun. 2014.
6. T. Agarwal and S. Cui, "Noncooperative games for autonomous consumer load balancing over smart grid," *Computing Research Repository - CoRR*, vol. abs/1104.3802, 2011. [Online]. Available: http://arxiv.org/abs/1104.3802
7. A.-H. Mohsenian-Rad, V. Wong, J. Jatskevich, R. Schober, and A. Leon-Garcia, "Autonomous demand-side management based on game-theoretic energy consumption scheduling for the future smart grid," *IEEE Transactions on Smart Grid*, vol. 1, no. 3, pp. 320–331, Dec. 2010.

8. I. E. A. (IEA), available online, accessed Aug 2015, "http://www.iea.org/statistics/".
9. Y. K. Tan, T. P. Huynh, and Z. Wang, "Smart personal sensor network control for energy saving in dc grid powered led lighting system," *IEEE Transactions on Smart Grid*, vol. 4, no. 2, pp. 669–676, Jun. 2013.
10. S. Arefifar, Y. Mohamed, and T. El-Fouly, "Supply-adequacy-based optimal construction of microgrids in smart distribution systems," *IEEE Transactions on Smart Grid*, vol. 3, no. 3, pp. 1491–1502, Sep. 2012.
11. D. Niyato and P. Wang, "Cooperative transmission for meter data collection in smart grid," *IEEE Communications Magazine*, vol. 50, no. 4, pp. 90–97, Apr. 2012.
12. L. Ochoa and G. Harrison, "Minimizing energy losses: Optimal accommodation and smart operation of renewable distributed generation," *IEEE Transactions on Power Systems*, vol. 26, no. 1, pp. 198–205, Feb. 2011.
13. M. Erol-Kantarci, B. Kantarci, and H. Mouftah, "Cost-aware smart microgrid network design for a sustainable smart grid," in *2011 IEEE GLOBECOM Workshops (GC Wkshps)*, Dec. 2011, pp. 1178–1182.
14. I.-K. Song, W.-W. Jung, J.-Y. Kim, S.-Y. Yun, J.-H. Choi, and S.-J. Ahn, "Operation schemes of smart distribution networks with distributed energy resources for loss reduction and service restoration," *IEEE Transactions on Smart Grid*, vol. 4, no. 1, pp. 367–374, Mar. 2013.
15. J. Aguero, "Improving the efficiency of power distribution systems through technical and non-technical losses reduction," in *Transmission and Distribution Conference and Exposition (T&D), 2012 IEEE PES*, Dallas, Texas, USA, May 2012, pp. 1–8.
16. T.-I. Choi and Y.-K. Cho, "International business offering related to innovative smart technologies," in *2012 IEEE PES Innovative Smart Grid Technologies (ISGT)*, Washington D.C., USA, Jan. 2012, pp. 1–8.
17. C. Wei, Z. M. Fadlullah, N. Kato, and A. Takeuchi, "GT-CFS: a game theoretic coalition formulation strategy for reducing power loss in micro grids," *IEEE Transactions on Parallel and Distributed Systems*, vol. 25, no. 9, pp. 2307–2317, Sep. 2014.
18. A. Abdallah and X. Shen, "Lightweight security and privacy preserving scheme for smart grid customer-side networks," *IEEE Transactions on Smart Grid*, to appear.
19. H. Li, X. Lin, H. Yang, X. Liang, R. Lu, and X. Shen, "EPPDR: an efficient privacy-preserving demand response scheme with adaptive key evolution in smart grid," *IEEE Transactions on Parallel and Distributed Systems*, vol. 25, no. 8, pp. 2053–2064, Aug. 2014.
20. M. M. Fouda, Z. M. Fadlullah, N. Kato, R. Lu, and X. Shen, "A lightweight message authentication scheme for smart grid communications," *IEEE Transactions on Smart Grid*, vol. 2, no. 4, pp. 675–685, Dec. 2011.
21. H. Zhu, X. Lin, R. Lu, P. Han Ho, and X. Shen, "SLAB: A secure localized authentication and billing scheme for wireless mesh networks," *IEEE Transactions on Wireless Communications*, vol. 7, no. 10, pp. 3858–3868, Oct. 2008.
22. X. Lin, R. Lu, P. Han Ho, X. Shen, and Z. Cao, "TUA: a novel compromise-resilient authentication architecture for wireless mesh networks," *IEEE Transactions on Wireless Communications*, vol. 7, no. 4, pp. 1389–1399, Apr. 2008.
23. IEEE P2030 draft guide for smart grid interoperability of energy technology and information technology operation with the electric power system (eps), and end-use applications and loads. [Online]. Available: http://grouper.ieee.org/groups/scc21/2030/2030_index.html
24. A. Hamlyn, H. Cheung, T. Mander, L. Wang, C. Yang, and R. Cheung, "Computer network security management and authentication of smart grids operations," in *2008 IEEE Power and Energy Society General Meeting - Conversion and Delivery of Electrical Energy in the 21st Century*, Jul. 2008, pp. 1–7.
25. G. N. Ericsson, "Cyber security and power system communication–essential parts of a smart grid infrastructure," *IEEE Transactions on Power Delivery*, vol. 25, no. 3, pp. 1501–1507, Apr. 2010.
26. A. R. Metke and R. L. Ekl, "Smart grid security technology," in *Innovative Smart Grid Technologies (ISGT)*, Jan. 2010, pp. 1–7.

27. K. Kursawe, G. Danezis, and M. Kohlweiss, "Privacy-friendly aggregation for the smart-grid," in *Proceedings of the 11th International Conference on Privacy Enhancing Technologies*, ser. PETS'11. Berlin, Heidelberg: Springer-Verlag, 2011, pp. 175–191. [Online]. Available: http://dl.acm.org/citation.cfm?id=2032162.2032172

28. M. Kgwadi and T. Kunz, "Securing RDS broadcast messages for smart grid applications," in *Proceedings of the 6th International Wireless Communications and Mobile Computing Conference*, ser. IWCMC'10. New York, NY, USA: ACM, 2010, pp. 1177–1181.

Chapter 4
Demand Response Challenge in Smart Grid

As indicated in Chap. 3, for the successful deployment of the smart grid, demand-side management (DSM) or demand response [1–3] is crucial. DSM means the art of planning and implementation of the electric utility activities that is aimed to influence the users' energy consumption by affecting desired changes in the shape of loads of the utility operator. While DSM aims at producing a change in the load-shape, it needs to balance the supply of the operator and the demand of the customers.

The traditional DSM methods either shift or reduce the energy consumption. Shifting the energy consumption can effectively reduce the aggregate energy load during the peak hours (which is the main cause of blackouts). To measure the imbalance in the load-shape of daily energy consumption, the ratio of the highest peak time energy consumption to the average consumption (PAR) [4] in an entire day is used. By shifting energy consumption from the peak hours to off-peak hours, PAR reduction is possible. Another DSM approach is to reduce the energy usage by encouraging energy-aware consumption patterns [5].

4.1 Background into DSM and Load Balancing Problem

Wang et al. [6] showed case studies of dynamic pricing programs, which were offered by some US electric companies. The case studies showed improvement of energy pricing performance through automated devices like smart thermostats to automatically decrease energy consumption of air-conditioning and central heating during peak hours. Automatic residential energy consumption scheduling was also used in [7] with an assumption that the advantages of real-time pricing are currently limited because currently buildings do not have good automation systems and customers feel difficult to adjust with dynamic energy pricing. The idea in [7] clearly shows the potential of using smart pricing in the smart grid. Samadi et al. [5] used

© The Author(s) 2015
Z.Md. Fadlullah, N. Kato, *Evolution of Smart Grids*, SpringerBriefs
in Electrical and Computer Engineering, DOI 10.1007/978-3-319-25391-6_4

a utility function in an energy schedule optimization in which power generation capacity over time is fixed. Samadi et al. [5] also used different energy pricing values for different users. But the customers felt it was unaware that they were given unequal pricing. To avoid the unfairness issue, the "load uncertainty" was added to the pricing scheme in [8]. Game theory was used to formulate the energy consumption scheduling problem in [9]. The objective of this game was to reduce PAR by shifting energy use. Each player (customer) plays the game with other players until the game reaches an equilibrium so that everyone becomes happy about their own PAR reduction. The game does have a distributed flavor, but typically it needs some central information. On the other hand, a purely centralized scheme for optimally scheduling energy use [4] was based on a social welfare function. This social welfare was actually the preference of the users to energy consumption, which is the difference of the utility and cost of the energy. However, utility of energy and cost of energy have different units and this was not entirely reasonable. In literature, we could find other centralized schemes such as [10] which use game-based interactions between energy companies, retailers, and customers. In the next section, let us overview an existing system model of smart grid for power, energy cost, and load control modeling [9].

4.2 System Model

We use the smart grid architecture shown in Fig. 2.1 as the basis of our adopted system model. In the adopted model, a single energy supplier (utility operator) having many customers is considered. The smart meters of the customers are connected with the operator's control center, and can schedule the energy consumption of their residence. Assume that the smart meters can monitor and collect the electricity consumption data of all electrical loads plugged into the grid. If required, the smart meters can also turn on or off the loads, and select the appropriate level of energy consumption for the loads. Furthermore, the smart meters can inform the operator regarding the energy consumption schedules of the customers. By this way, the smart meters could be assumed to be capable of informing the power company or the supplier about users' energy consumption schedules.

Our considered smart grid model has the following aspects, the power system modeling, appropriate energy cost function modeling, and load control on consumer-end modeling [3, 11].

4.2.1 Power System Modeling

We use a popular power system modeling as in [9] which supposes $\mathcal{N}$, a set of users, who obtain electricity from the utility operator. The number of users is $N \triangleq |\mathcal{N}|$. For each user, $n \in \mathcal{N}$, l_n^h denotes the total load during hour $h \in \mathcal{H} \triangleq \{1, \ldots, H\}$

where $H = 24$. The daily load of the nth user is given by her energy consumption vector, $\mathbf{l}_n \triangleq \{l_n^1, \ldots, l_n^H\}$. Let L_h denote the aggregate load of the users during each hour of a day (i.e., $h \in \mathcal{H}$). This can be calculated using $L_h \triangleq \sum_{n \in \mathcal{N}} l_n^h$. The daily peak and average load levels are calculated as $L_{peak} = \max_{h \in \mathcal{H}} L_h$ and $L_{avg} = \frac{1}{H} \sum_{h \in \mathcal{H}} L_h$, respectively. Now we can calculate the PAR $= \frac{L_{peak}}{L_{avg}} = \frac{H \max_{h \in \mathcal{H}} L_h}{\sum_{h \in \mathcal{H}} L_h}$.

4.2.1.1 Modeling an Appropriate Energy Cost Function

Consider that the hourly energy price changes in the model. The change is proportional to the system energy consumption in that hour. This is done so as to give incentive to the customers to prevent them from using much energy during peak hours. Such a policy can contribute to a lower PAR. The cost function is an increasing one which means that the energy cost increases with L_h (the total energy consumption). For each $h \in \mathcal{H}$, we now have $C_h(L_h^1) < C_h(L_h^2), \forall L_h^1 < L_h^2$. This assumption makes sure that the higher the energy consumption, the higher the impact on the energy price hike. Also, the energy cost functions are assumed to be strictly convex such that for every $h \in \mathcal{H}$, $C_h(\theta L_h^1 + (1 - \theta)L_h^2) < \theta C_h(L_h^1) + (1 - \theta)C_h(L_h^2)$. Here, L_h^1, L_h^2, and θ are real numbers such that $L_h^1, L_h^2 \geq 0$ and $0 < \theta < 1$.

4.2.1.2 Modeling Load Control on the Customer-Side

Consider A_n to be the set of loads (including air-conditioners, heaters, refrigerators, etc.) of the $n^{th} \in \mathcal{N}$ customer. For each of her load, an energy consumption scheduling vector $x_{n,a} = [x_{n,a}^1, \ldots, x_{n,a}^H]$ is constructed where $x_{n,a}^h$ means the corresponding one-hour energy consumption scheduled for the load a during hour h. Suppose $l_n^h = \sum_{a \in A_n} x_{n,a}^h, h \in H$ denotes her total energy consumption, $L_h = \sum_{n \in N} l_n^h, h \in H$ means the aggregate energy consumption of all the users, and $L_{-n,h} = L_h - l_n^h, h \in H$ indicates the aggregate energy consumption of all users except the nth customer. Then, the smart meter of each customer needs to decide the optimal schedule of energy consumption vector, $x_{n,a}$, for all her loads with some constraints like starting and stopping time of the schedule ($\alpha_{n,a} \in \mathcal{H}$ and $\beta_{n,a} \in \mathcal{H}$, respectively), and the minimum and maximum energy requirement ($E_{n,a}^{min}$ and $E_{n,a}^{max}$, respectively) to finish the operation. Then, the time interval within which the loads may be scheduled is given by $\mathcal{H}_{n,a}$. The power level of each load, $a \in A_n$, also is limited by the minimum standby power level $\gamma_{n,a}^{min}$, and the maximum power level $\gamma_{n,a}^{max}$. Then, we have: $\gamma_{n,a}^{min} \leq$ & $x_{n,a}^h \leq \gamma_{n,a}^{max}, \forall h \in \mathcal{H}_{n,a}$ and $x_{n,a}^h = 0, \forall h \in \mathcal{H} \backslash \mathcal{H}_{n,a}$. The feasible energy consumption scheduling set of the nth customer is then derived to be:

$$\chi_n = \{x_n | E_{n,a}^{min} \leq \sum_{\alpha_{n,a}}^{\beta_{n,a}} x_{n,a}^h \leq E_{n,a}^{max},$$

$$\gamma_{n,a}^{min} \leq x_{n,a}^h \leq \gamma_{n,a}^{max}, \quad \forall h \in \mathcal{H}_{n,a},$$

$$x_{n,a}^h = 0, \quad \forall h \in \mathcal{H}\backslash\mathcal{H}_{n,a}\}. \tag{4.1}$$

In order to improve the assumptions of the existing model, we incorporate a number of modifications to it by adopting specific strategies of the users and the utility operator, and formulate a more complete optimization problem.

4.3 Adopted Strategies of Users and Utility Operator and a Game-Theoretic Solution

In the following, we first discuss our adopted strategies of the users and utility operators by making some modifications to the earlier described existing system model described in the earlier section. Based on these strategies, a two-step game is then presented.

4.3.1 Strategy of the Users

Instead of optimizing the energy consumption schedules of all the users at once in (4.1), each smart meter needs to optimize the schedule of its user based on her energy need. Obviously her energy consumption schedule vector belongs to the feasible set in (4.1). Since the objective is to optimize the users' pay-off, suppose that the objective function is the following function of energy consumption schedule vector.

$$W_i(x_1, \ldots, x_{24}) = Value\ of\ Energy - Cost\ of\ Energy$$

$$= V(X_i) - \sum_{h=1}^{24} x_h C_h(L_h), \tag{4.2}$$

where $(x_1, \ldots, x_{24})$ is the energy consumption scheduling vector with the energy loads of user i during the hours of the day, $X_i = \sum_{h=1}^{24} x_h$ means the total energy consumption, and $V(X_i)$ represents the value of that amount of energy.

In the remainder of the section, we present an analysis of the value of energy function and cost of energy function.

4.3.1.1   Value of Energy Perceived by Users

Let $V(X_i)$ indicate how much energy value is given by the user i. Because each user in the power system could have different energy consumption patterns, their energy consumption schedules are likely to be different also. Even in the case where some users have the same energy consumption, their attitude and habits may dictate their perceived energy value. So, it is not so trivial to capture the response and energy demand of different users toward the same energy price. Hence, we analytically model the users' preferences toward energy consumption by adopting the utility function theory of microeconomics [12].

According to utility theory, a reasonable utility function should satisfy the characteristics of quadratic and exponential functions. From [13], the exponential function could present a better model of the user's preference which motivates us to adopt the exponential function $u(X) = 1 - e^{-\omega X}$ as its value domain has been normalized between (0,1). Figure 4.1 shows a plot of utility functions in which ω is a parameter of the users' tolerance toward energy consumption reduction. The figure demonstrates that a utility function with a relatively larger value of ω slowly approaches the upper-bound. This means that the users with large ω values are strict with the energy curtailment. In other words, their utility values are lower compared to other users even with the same energy consumption. Because the ω value of a user is her private information, it should not be revealed to other users. Based on this utility function, we define the value function of energy as:

$$V(X_i) = pX_{max}u(X_i) = pX_{max}(1 - e^{-\omega X_i}). \tag{4.3}$$

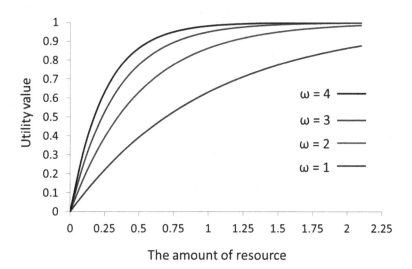

Fig. 4.1 Value of utility for different resources

In (4.3), X_{max} is the maximum amount of energy that the user can use. p is the average price of a unit of energy.

The adopted utility function has the maximum value of 1. By scaling it up with the product of average energy price and the maximum amount of energy consumption of the user, the maximum value of $V(X_i)$ becomes equal to the value of the maximum energy consumption. What this means is that the user will value her consumption amount not over the average money she needs to pay to satisfy her maximum demand.

4.3.1.2   Cost of Energy

Compared to the piece-wise and quadratic linear functions described in [4, 9], the gap in energy price between peak hours and off-peak hours is large. For example, PAR $= 3$ means that the load during peak-hours is three times the average daily load. This indicates that the difference between the prices of energy is significant. Due to strict schedules during peak hours, this large difference is inconvenient to the users. Instead of making such a big jump in the pricing, a proportional increase in the energy cost in accordance with the total load is important to encourage participation of users in balancing the PAR. In [6], the energy prices were carefully selected by considering the reaction of the users and not allowing the differences between energy prices to exceed three times.

We could intuitively consider that the more drastically the cost function changes, the better PAR reduction we would get. However, the computation time increases much when the energy cost function varies dramatically. So, there is actually a trade-off in selecting the energy price. We adopt the following energy price function with this trade-off in mind.

$$C_h(L_h) = \alpha L_h \log(L_h + 1), \tag{4.4}$$

where α is our "price parameter." The utility operator can tune α to change the daily energy price to control the energy consumption of the users. The price difference between the peak and off-peak hours still remains unchanged. The logarithmic function in (4.4) gives almost a linear shape (Fig. 4.2). The comparison between our adopted price function and the conventional quadratic price function can also be noticed in Fig. 4.2.

By substituting (4.3) and (4.4) into (4.2), the user i's objective function can be rewritten as:

$$W_i(x_1, \ldots, x_{24}) = pX_{max}(1 - e^{-\omega \sum_{h=1}^{24} x_h})$$

$$- \sum_{h=1}^{24} \alpha x_h L_h \log(L_h + 1). \tag{4.5}$$

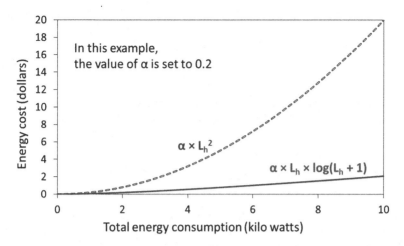

Fig. 4.2 The comparative increase of energy price with the total energy consumption using the quadratic function and the extended energy price function

4.3.2 Strategy of the Utility Operator

Now we consider how the utility operator determines the energy pricing scheme. In our enhanced model, the energy price varies according to the total load during different hours. For the newly adopted price function, the energy price is now proportional to $\big(L_h \log(L_h+1)\big)$. This is ensured by setting the energy price function as $\big(C_h(L_h) = \alpha L_h \log(L_h + 1)\big)$, where L_h is the total load during hour h. Thus, we have an energy price vector, $C(\mathcal{L}) = \alpha \cdot \big(L_1 \log(L_1+1)+\ldots+L_{24} \log(L_{24}+1)\big)$, for the entire day. Therefore, the supplier can tune the parameter α so as to influence the price vector, and thus, limit the users' energy consumption. How can the operator decide a reasonable α value? Consider a baseline fixed price scheme used in traditional power grids. In Japan, TEPCO [14] charges 17.87 yen for the first 120 kW (kilowatts) while 22.86 yen for the next 180 kW for the light residential user with 20 ampere power line. Hence, assume that the operator has the average price for the energy which it sells to the users. This average price was actually the parameter p in (4.3). If the operator uses the dynamic price scheme, assume that the total cost charged for the entire system is equal to the fixed price scheme with the same load. With this relaxed assumption, we have:

$$p \sum_{h=1}^{24} L_h = \sum_{h=1}^{24} L_h C_h(L_h)$$

$$= \alpha \sum_{h=1}^{24} L_h^2 \log(L_h + 1). \qquad (4.6)$$

Then, the operator can calculate the energy price parameter α as follows.

$$\alpha = \frac{p \sum_{h=1}^{24} L_h}{\sum_{h=1}^{24} L_h^2 log(L_h + 1)}. \tag{4.7}$$

Now that we have all the necessary parameters, we are set to show how a demand game can be played between the operator and its users.

4.4 Playing the Demand Game

The game played between the utility operator and users aims to reduce the system PAR by optimizing energy schedules of the customers[11]. The optimization process can be modeled as a two-stage game [15] as shown in Algorithms 1 and 2.

1. The users will try to maximize their pay-offs by optimizing functions (shown in (4.5)) using interior point method (IPM).
2. The supplier will then adjust the energy price parameter consistent with the user's energy consumption schedule according to (4.7).

When the game reaches an equilibrium state, neither users nor supplier will change their strategies. At the same time, the system PAR and total energy consumption are reduced.

Algorithm 1 Utility operator's game.

 1: **Begin**: Collect original schedules from all users
 2:     All users initialize their schedules from feasible sets
 3: **End**
 4: Calculate initial energy price parameter α according to (4.7)
 5: **Repeat**
 6:     Randomly choose user $n \in N$
 7:     Signal user n to execute **algorithm 2**
 8:     Update the new schedule vector $\mathcal{L}_n$ from user n
 9:     Update the energy price parameter α according to (4.7)
10: **Until** No user wants to change schedule

Algorithm 2 User's game.

 1: **Begin**: Receive signal from operator
 2:     Request α, vector $\mathcal{L}$
 3:     User n optimizes schedule by solving the problem in (4.5) by using IPM
 4:     **If** x_n changes compared to current schedule **Then**
 5:         Inform the center of the new schedule vector $\mathcal{L}_n$
 6:     **End If**
 7: **End**

4.5   Comparative Performance

To demonstrate the effectiveness of our adopted game-theoretic approach, its performance is compared with two conventional methods (centralized convex optimization and energy consumption game [9] which are referred to as conventional 1 and conventional 2 for easy reference). Using MATLAB [16], a smart grid system having a single utility operator with multiple users was modeled. The number of users was varied from 5 to 500. Each user is assumed to be connected to the operator's control center via smart meter. Each user is supposed to have 10–15 schedulable loads (washing machines, dish washers, PHEVs, etc.) and 10–15 non-schedulable loads (lighting banks, heaters, etc.). Table 4.1 lists energy consumption of some typical residential loads [17]. The settings are randomly generated during each simulation run. However, the same setting is maintained in a specific situation for comparing the adopted approach with the conventional methods.

Figure 4.3 shows that the running time of conventional 1 increases rapidly (almost at an exponential rate) whereas the two game-theoretic methods (the adopted approach and conventional 2) require low completion time with increasing numbers of users. Even with a relatively low number of users, the large number

Table 4.1 Energy appliances and their average consumption on a daily basis

Appliances	Average consumption per day (kW)
Clothes dryer	2.47
Dishwasher	0.99
Lighting	3.29
Refrigerator	5.89
Washing machine	0.28

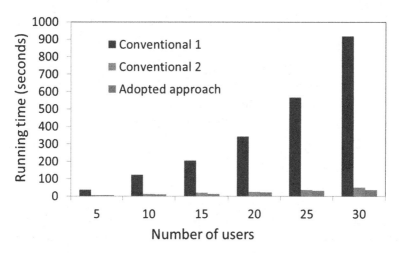

Fig. 4.3 Comparison of the running time between the conventional methods and the adopted approach

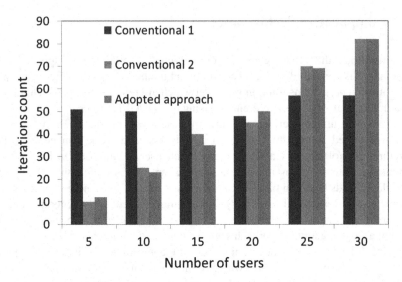

Fig. 4.4 Number of iterations until convergence in case of the conventional methods and the adopted approach, for varying numbers of users

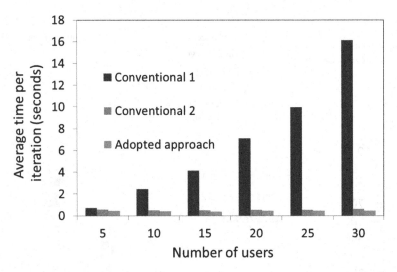

Fig. 4.5 Average time needed for each iteration in case of the conventional methods and the adopted approach

of parameters seriously influences the running time of convex optimization (i.e., conventional 1). Also, it may not lead to convergence for users exceeding 30. So, conventional 1 seems not to be viable. Figures 4.4 and 4.5 will shed more light into this issue.

In Figs. 4.4 and 4.5, the number of iterations required for each algorithm to converge and the average time per iteration are demonstrated, respectively. The running time in the previous analysis is the outcome of these two parameters. Figure 4.4 demonstrates how the number of iterations needed for conventional 1, conventional 2, and our adopted approach gradually increases with the number of users. The average time required for each iteration, however, as demonstrated in Fig. 4.5, varies slowly for the game-theoretic algorithms while it rises drastically in case of conventional 1. This corroborates our expectation because the game-theoretic approaches need to solve just the local optimization problem of each user. These indicate that conventional 1 is not scalable enough for a large number of users whereas the game-theoretic approaches perform much better. In the next comparisons, we only compare conventional 2 and the adopted approach since conventional 1 was already shown not to be scalable with the growing number of users.

Figure 4.6 depicts the number of iterations required for the convergences of conventional 2 and the adopted approach. Their convergence is fast because both are based on game theory. Also with increasing numbers of users, the ratio between the number of iterations and the number of users decreases a little. This means that the bigger the system becomes, the less effect is inflicted by changing a single user's schedule. Figure 4.7 demonstrates the comparison of the PAR reduction in conventional 1 and our adopted approach. The results indicate that the latter achieve much higher PAR reduction because of its consideration of not only shifting energy consumption while scheduling, but also its ability to adjust energy consumption levels during different hours.

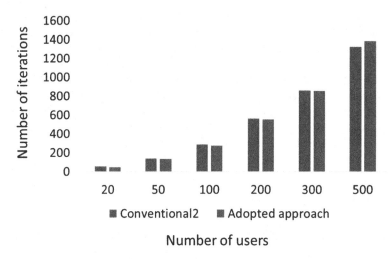

Fig. 4.6 Number of iterations required for the conventional 2 method and the adopted approach for a large number of users (up to 500)

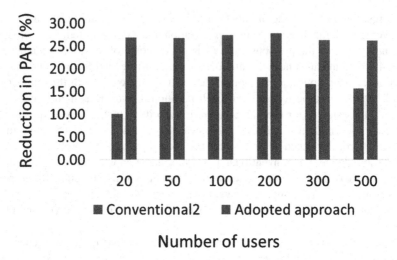

Fig. 4.7 PAR reduction in the conventional 2 method and the adopted approach for varying numbers of users

4.6 Concluding Remarks

In this chapter, a smart grid infrastructure was considered where the utility operator and its consumers were demonstrated to play their respective games to optimize their energy schedules. In the adopted two-stage game in smart grid, the objective was to achieve a reduction in the system PAR. A pricing scheme was implemented in which the energy price changes according to the total energy consumption during each hour so that the customers receive reasonable incentives to follow the system. Because many parameters need to be optimized, a fully centralized control using convex optimization did not perform well. The comparative performance also demonstrated that both the conventional and the adopted approaches which were based on game theory converged much faster and scaled well with the growing number of users. In particular, our adopted approach achieved much better PAR reduction which is desired from the utility operator's perspective. At the same time, the users could reduce their energy cost in an effective way.

References

1. M. M. Fouda, Z. M. Fadlullah, N. Kato, A. Takeuchi, and Y. Nozaki, "A novel demand control policy for improving quality of power usage in smart grid," in *2012 IEEE Global Communications Conference (GLOBECOM)*, Anaheim, California, USA, Dec. 2012, pp. 5154–5159.
2. C. Gellings, "The concept of demand-side management for electric utilities," *Proceedings of the IEEE*, vol. 73, no. 10, pp. 1468–1470, Oct. 1985.

3. Z. M. Fadlullah, M. Q. Duong, N. Kato, and I. Stojmenovic, "A novel game-based demand side management scheme for smart grid," in *2013 IEEE Wireless Communications and Networking Conference (WCNC)*, Shanghai, China, Apr. 2013, pp. 4677–4682.

4. P. Samadi, R. Schober, and V. W. S. Wong, "Optimal energy consumption scheduling using mechanism design for the future smart grid," in *2011 IEEE International Conference on Smart Grid Communications (SmartGridComm)*, Brussels, Belgium, Oct. 2011, pp. 369–374.

5. P. Samadi, A.-H. Mohsenian-Rad, R. Schober, V. Wong, and J. Jatskevich, "Optimal real-time pricing algorithm based on utility maximization for smart grid," in *2010 First IEEE International Conference on Smart Grid Communications (SmartGridComm)*, Gaithersburg, Maryland, USA, Oct. 2010, pp. 415–420.

6. J. Wang, M. Biviji, and W. Wang, "Lessons learned from smart grid enabled pricing programs," in *2011 IEEE Power and Energy Conference at Illinois (PECI)*, Urbana, IL, USA, Feb. 2011, pp. 1–7.

7. A.-H. Mohsenian-Rad and A. Leon-Garcia, "Optimal residential load control with price prediction in real-time electricity pricing environments," *IEEE Transactions on Smart Grid*, vol. 1, no. 2, pp. 120–133, Sep. 2010.

8. P. Tarasak, "Optimal real-time pricing under load uncertainty based on utility maximization for smart grid," in *Smart Grid Communications (SmartGridComm), 2011 IEEE International Conference on*, Brussels, Belgium, Oct. 2011, pp. 321–326.

9. A.-H. Mohsenian-Rad, V. Wong, J. Jatskevich, R. Schober, and A. Leon-Garcia, "Autonomous demand-side management based on game-theoretic energy consumption scheduling for the future smart grid," *IEEE Transactions on Smart Grid*, vol. 1, no. 3, pp. 320–331, Dec. 2010.

10. S. Bu, F. Yu, and P. Liu, "Dynamic pricing for demand-side management in the smart grid," in *Online Conference on Green Communications (GreenCom), 2011 IEEE*, Sep. 2011, pp. 47–51.

11. Z. M. Fadlullah, D. M. Quan, N. Kato, and I. Stojmenovic, "GTES: an optimized game-theoretic demand-side management scheme for smart grid," *IEEE Systems Journal*, vol. 8, no. 2, pp. 588–597, Jun. 2014.

12. "An introduction to utility theory," accessed Aug. 2015. [Online]. Available: http://www.norstad.org/finance/util.pdf

13. G. H. and P. G., "Utility functions: From risk theory to finance," *North American Actuarial Journal (NAAJ)*, vol. 2, no. 3, pp. 74–100, 1998.

14. "TEPCO," accessed Aug. 2015. [Online]. Available: http://www.tepco.co.jp/e-rates/individual/data/

15. J. Zhang and Q. Zhang, "Stackelberg game for utility-based cooperative cognitive radio networks," in *Proceedings of the Tenth ACM International Symposium on Mobile Ad Hoc Networking and Computing*, ser. MobiHoc '09. New York, NY, USA: ACM, 2009, pp. 23–32. [Online]. Available: http://doi.acm.org/10.1145/1530748.1530753

16. Mathworks - matlab and simulink for technical computing. [Online]. Available: http://www.mathworks.com/

17. "Typical power consumption," accessed Aug. 2015. [Online]. Available: http://www.oksolar.com/technical/consumption.html

Chapter 5
Game-Theoretic Coalition Formulation Strategy for Reducing Power Loss in Micro Grids

5.1 Background

In this chapter, we move away from the smart-grid demand-side management topic to another important one of micro grids. Even though a significant progress has been achieved in the development of the micro grids, the power loss minimization between the micro grids and also between the macro station and an individual micro grid is still receiving much attention. Niyato and Wang [1] considered an algorithm aimed to optimize the transmission strategy in order to minimize the total cost including the power loss. The power losses minimization in the energy distribution networks has conventionally been investigated using a single and deterministic demand level. A novel algorithm to overcome this problem was designed in [2]. The "cost-aware smart micro grid network design" in [3] allows economic power transactions within the smart grid with manageable power losses. Meliopoulos et al. [4] also discussed power loss minimization issues by proposing a coordinated control scheme at real-time with the inclusion of distributed generation resources (micro grids) with the existing grid. A novel load management solution to coordinate the charging of multiple PHEVs in a smart grid system was considered in [5] which also considered the power loss problem. An efficient optimal reconfiguration algorithm for power loss minimization was studied in [6]. Costabeber et al. [7] demonstrated that the power loss reduction is viable without central controllers by exploiting local measurement, communication, and control capability in the micro grids. Saad et al. [8] used cooperative game theory to formulate novel cooperative strategies between the micro grids of an energy distribution network aimed to reduce the power loss. Also, Costabeber et al. [9] and Tenti et al. [10] discussed cooperative operation of neighboring power processing units to reduce distribution losses. Corso et al. [11] addressed the daily schedule of distributed generators to minimize power loss in a micro grid connected to the main grid. A more detailed methodology to develop an autonomous micro grid for coping with power loss can be found in [12]. Also, a heuristic-based greedy algorithm was proposed

© The Author(s) 2015
Z.Md. Fadlullah, N. Kato, *Evolution of Smart Grids*, SpringerBriefs
in Electrical and Computer Engineering, DOI 10.1007/978-3-319-25391-6_5

in [13] to reduce the power loss. A review of technologies, methodologies, and operational approaches for improving the efficiency of power distribution systems like micro grids can be accessed in [14]. The "autonomous regional active network management system" project aimed to design a distribution network power loss management algorithm in [15]. The tie-set graph theory and its application to smart grid networks was introduced to minimize the power loss in [16]. Power loss of the energy distribution networks was also considered in [17–21].

Compared to the aforementioned studies addressing the power loss minimization problem of distribution networks in a scattered manner, we need to have a more comprehensive insight into this issue in the micro grids context. This chapter will describe a basic micro grids system model and illustrate how formulation of effective coalitions between the micro grids can be useful in significantly reducing the distribution power loss.

5.2 System Model

Figure 5.1 shows our considered system model comprising a number of micro grids.

In this section, our considered system model of the smart grid is presented. This model considers that the users are supplied electricity by the macro station and/or a number of autonomous micro grids using distributed, renewable energy sources like wind farms, solar panels, PHEV batteries, etc. Each micro grid is linked

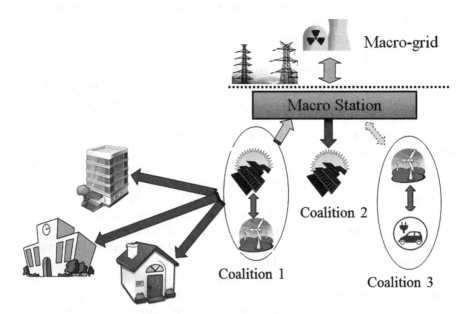

Fig. 5.1 Example of micro grids oriented system model

to the macro station through the main power grid. Also, each micro grid is assumed to have its own customers such as residential users, schools, factories, etc., who have the capability to notify their respective micro grid(s) regarding energy demands via AMI. Because the power loss between the macro station and the micro grid is usually more than that between two neighboring micro grids, the adjacent micro grids could potentially improve the power loss through forming supportive groups, which we refer to as coalitions. A coalition can initially form with just one micro grid (e.g., coalition 2 in Fig. 5.1).

Let the set of micro grids be $\mathcal{N}$. Assume that during each hour, the ith micro grid (for $\forall i \in \mathcal{N}$) denoted by MG_i produces energy G_i and confronts the total demand D_i from all its users. Let $Req_i = (G_i - D_i)$ be the surplus of MG_i. If $Req_i < 0$, MG_i needs to buy additional energy to meet the total demand of its users. On the other hand, MG_i has energy surplus and can sell its energy if $Req_i > 0$. If $Req_i = 0$, MG_i just meets the demand, and in this case, it is neither an energy seller nor a buyer. Based on this concept, all the micro grids can be divided into three types—neutral, seller, or buyer.

The production G_i and the total demand D_i are supposed as random numbers in the real Smart Grid networks [22] since D_i depends on many unpredictable factors like time, user behavior, etc. Furthermore, because many micro grids use renewable resources, their energy generation is subject to surrounding environment, G_i cannot be treated as a constant either. So, Req_i is also a random number having a certain observed distribution.

Additionally, assume that a micro grid acting as seller has enough energy to sell to the micro grids who assumed the role of buyer(s). On the other hand, assume that a buyer has sufficient "money" to buy the energy from the seller(s) and/or the macro station.

Now we can extend the concept of buyers and sellers to the groups or coalitions formed by a number of micro grids. For instance, Fig. 5.1 shows that there are three coalitions. In the considered time window (during the current hour), coalition 1 has energy surplus and can sell energy to other micro grids or the macro station. But, coalition 2 needs energy and will need to buy energy from another coalition or multiple other coalitions and/or the macro station. On the other hand, Coalition 3's demand and supply are equal, so the micro grids in this coalition do not need to exchange energy with any other coalition or the macro station.

5.2.1 Non-cooperative Model

Before we can establish the payoff function of coalition, consider a non-cooperative case as the baseline/comparable model where each micro grid only sells energy to the macro station or buys energy from the macro station. Let U_0 express the medium level voltage of power transfer between a micro grid and the macro station. Any energy transfer between the micro grid and the macro station will have power loss. In this chapter, we are concerned with the basic power losses related to distribution, namely the power loss over the distribution lines inside the network, and the power loss due to other factors in the macro station such as friction, corrosion, weather, dust, etc. If MG_i wants to sell Req_i to the macro station when $(Req_i > 0)$, or buy Req_i from the macro station when $(Req_i < 0)$, the power loss P_{i0} can be expressed as:

$$P_{i0} = R_{i0}I_0^2 + \alpha Q_i, \tag{5.1}$$

where R_{i0} denotes the distribution line resistance between the macro station and MG_i, and α is a constant to account for a fraction of power loss due to other factors. Q_i denotes the energy which MG_i wants to buy or sell. $I_0 = \frac{Q_i}{U_0}$ represents the current flowing over the distribution line when there is an energy transfer between the macro station and MG_i. Then, we can rewrite (5.1) as:

$$P_{i0} = \frac{R_{i0}Q_i^2}{U_0^2} + \alpha Q_i, \tag{5.2}$$

where

$$Q_i = \begin{cases} Req_i & : & Req_i > 0 \\ L_i^* & : & Req_i < 0 \\ 0 & : & Req_i = 0. \end{cases} \tag{5.3}$$

Here, L_i^* is the total amount of power which needs to be generated (or be made available to the system) to make sure that MG_i is able to get the energy needed to meet its total demand Req_i. If there is no power loss, $L_i^* = |Req_i|$. Hence, $L_i^* > |Req_i|$. L_i^* is the solution of following quadratic equation.

$$L_i = P_{i0} + |Req_i| = \frac{R_{i0}L_i^2}{U_0^2} + \alpha L_i - Req_i. \tag{5.4}$$

The number of roots of (5.4) depends on the values of the resistance R_{i0}, transfer voltage U_0, and α. For a specific Req_i value, three possible solution-sets of (5.4) exist (none, one, and two solutions). Because we are interested in minimizing the value of Req_i, if (5.4) has two roots, the smaller one is to be used. If (5.4) has no solution, assume that the root is the same as Eq. (5.4) having a single root, which is $L_i^* = \frac{(1-\alpha)U_0^2}{2R_{i0}}$.

As expressed by (5.2) and (5.3), the power loss between MG_i and the macro station depends on several factors such as the resistance R_{i0}, transfer voltage U_0, α, and Req_i (for either buying or selling). Hence, we can estimate the payoffs of the micro grids, when the parameters are given. Because in the non-cooperative case each micro grid can be considered to form a coalition, the payoff of the micro grid is equal to that of the coalition. So, we can define the non-cooperative payoff (utility) of each MG_i as the total power loss due to the power transfer:

$$u(\{i\}) = -w_2 P_{i0}, \tag{5.5}$$

where w_2 is the price of a unit power in the macro station. Since we aim to minimize $u(\{i\})$, the negative sign is able to convert the problem into a problem of seeking the maximum.

5.2.2 Cooperative Coalition Model

Now consider the cooperative coalition model for managing the micro grids acting as buyers and sellers. In addition to exchanging energy with the macro station, the micro grids are able to exchange energy with each other. This is particularly attractive to the micro grids because the power loss during transmission between the neighboring micro grids is always less than that between the macro station and a distant micro grid. Furthermore, the micro grids can make collaborative groups or coalitions as explained earlier to exchange energy with each other so as to minimize the power loss in the main smart grid and maximize their payoffs in (5.5).

To study the cooperative behavior of the micro grids, consider the framework of coalition game theory in [22] which defines a coalition game to have a set of players $\mathcal{N}$, the strategy of players, and the function $v: 2^{\mathcal{N}} \to \mathbb{R}$. In this game, v denotes a function which assigns to every coalition $S \subseteq \mathcal{N}$ a real number representing the total profits achieved by S. Any coalition $S \subseteq \mathcal{N}$ can be divided into two parts: the set of sellers ($S_s \subset S$) and the set of buyers ($S_b \subset S$) such that $S_s \cup S_b = S$. So, for a micro grid $MG_i \in S_s$, $Req_i > 0$ and it implies that MG_i can sell energy to others. On the other hand, a micro grid $MG_j \in S_b$ with $Req_j < 0$ implies that it wants to buy energy from others. Obviously, any coalition $S \subseteq \mathcal{N}$ needs to have at least one seller and one buyer.

To calculate the payoffs of all the coalitions, let us define the payoff function $v(S)$ for each $S \subseteq \mathcal{N}$. For any coalition $S = S_s \cup S_b$, we need to study the local power transfer between the sellers S_s, the buyers S_b, and the macro station. Consider how to form the coalition. Note that a micro grid does not join a coalition, if the payoff of this micro grid in the coalition is below than the payoff out of the coalition. Also, if a micro grid cannot give payoff to the coalition, it will not join that coalition.

Unlike the non-cooperative model, in the formed coalition in the cooperative model, there could be multiple micro grids, which can exchange energy with others or even with the macro station. Let $MG_i \in S_s$ and $MG_j \in S_b$ be a seller and a buyer, respectively. When MG_i and MG_j want to exchange power, the power loss function P_{ij} is:

$$P_{ij} = \frac{R_{ij}Q_{ij}^2}{U_1^2}, \tag{5.6}$$

where R_{ij} denotes the resistance of the distribution line between MG_i and MG_j. U_1($<$ U_0) is the transfer voltage between MG_i and MG_j. When $\alpha = 0$, (5.6) becomes the special case of (5.2). Also, $Q_{ij} = min(Q_i, Q_j)$ where Q_i and Q_j are given by (5.3).

It means that if the seller (MG_i) cannot meet the demand of the buyer (MG_j), then the seller only sells Q_i to the buyer. Also, after accounting for the negligible yet existing power loss between the micro grids, MG_j will buy at least $\frac{U_i^2}{2R_{ij}}$ amount of energy (due to the power loss between MG_i and MG_j) from MG_i.

In any coalition S, the total payoff function consists of three parts: the power loss between the micro grids which can be obtained from (6.13), the power loss caused by the micro grid selling energy to the macro station, and the power loss caused by the micro grid buying energy from the macro station. The second and third parts are given by (5.2) and (7.7), respectively. Hence, the total payoff function of the coalition S can be expressed as:

$$u(S, \Omega) = -(w_1 \sum_{i \in S_s, j \in S_b} P_{ij} + w_2 \sum_{i \in S_s} P_{i0} + w_2 \sum_{j \in S_b} P_{j0}), \tag{5.7}$$

where $\Omega \in \mathcal{S}_S$ denotes the join order of the micro grids, which decide to join the coalition S, and $\mathcal{S}_S$ indicates the set of the micro grids' order in S. w_1 and w_2 denote the price of a unit power in the coalition and that in the macro station, respectively. P_{i0} and P_{j0} are given by (5.2) and (7.7), respectively. They represent the power losses during the energy transfer between the corresponding micro grids and the macro station. P_{ij} indicates the power loss inside the coalition S, between $MG_i \in S_s$ and $MG_j \in S_b$ which are also expressed by (6.13). Using (5.7), which represents the total power loss due to the different power transfers for S, we can define the value function for the micro grids $(\mathcal{N}, v)$ coalition game:

$$v(S) = \max_{\Omega \in \mathcal{S}_S} u(S, \Omega) \tag{5.8}$$

5.3 Coalition Formation Game

In a game, the players (participating micro grids in our context) have a number of options that they can choose from [23]. For example, which coalitions to join, which coalition/micro grids/macro station to buy the power from, etc. Each player aims to select the best possible choice. The price of electricity from the micro grid is assumed to be cheaper than that from the macro station during peak hours and the power loss between the micro grids is supposed to be less than that between the micro grid and the macro station. So, an efficient strategy is needed to make sure that the aggregate energy transfers between the coalitions and the macro station is the minimum or the aggregate energy transfers within the coalitions is the maximum. This is to maximize the payoffs of the micro grids. Remember from (5.7) that $u(S, \Omega)$ has three parts. Because the power loss between the macro station and micro grids is much higher than that between the neighboring micro grids forming a coalition, the first part is much lower than the second part. Also, similar observation can be made about the third part. Hence, to maximize (5.8), a strategy is required to find the coalition with the appropriate micro grids so as to ascertain the minimum power loss between the coalition and the macro station, and the maximum power within the coalition. Our adopted strategy is as follows.

- Initialization: order S_s and S_b in accordance with the requests of the seller/buyer micro grids, i.e. $S_b = \{b_1, \ldots, b_k\}$, and calculate the sum of the sets respectively and find the least one of them. Assume S_b to be the least one. Then, choose $b_l \in S_b$ as the objective.
- Step 1: depending on the demand, find the *appropriate* micro grids in S_s or S_b to form coalition S with objective. This ensures that the power loss of coalition S is the minimum in the set of micro grids coalitions and the power exchanged in the groups is the maximum.
- Step 2: If the remainder of S_s is less than that of S_b, select the largest one in S_s as the objective. Go to step 1 until there is no availability in the sets or one of the sets becomes empty.
- Step 3: If the remainder of S_s is more than that of S_b, select the largest one in S_b as the objective. Go to step 1 until there is no availability in the sets or one of sets becomes empty.

Revisit (5.8), which shows the maximum total utility produced by any $S \in \mathcal{N}$ with the minimum power loss over the distribution lines. Therefore, comparing with the non-cooperative case, the sum of utilities of the micro grids in the cooperative coalition model increases. In other words, the micro grids receive extra profits by forming the coalition. After making the coalitions, the micro grids now face a new problem—how to distribute the extra profits appropriately in the coalition? This is important for stability because inappropriate allocation of the extra profits will lead to split of the coalition. The "Shapley" value concept from cooperative game theory can be exploited for this purpose that assigns a unique distribution (among the micro grids) of a total surplus produced by the coalition of all the micro grids. In other words, the Shapley value of a micro grid is simply the contribution of that micro grid to its coalition. If there exists a coalition game $(\mathcal{N}, v)$, the Shapley value can be calculated using:

$$\phi_i(v) = \sum_{S \subseteq \mathcal{N} \setminus \{i\}} \frac{|S|!(n - |S| - 1)}{n!} (v(S \cup \{i\}) - v(S)), \qquad (5.9)$$

where n denotes the total number of players (i.e., micro grids) and the sum extends over all the subsets S of $\mathcal{N}$ without the ith player. Imagine that the coalition is constructed with one player at a time; then each player demands her contribution $v(S \cup \{i\}) - v(S)$ as an appropriate compensation, and then averages over the possible different permutations in which the coalition may be formed. Because the payoffs depend on the micro grids' order in the coalition, the payoffs are expected to be different in different orders. Therefore, the contribution of a micro grid to the coalition is actually independent of the order. So, the fraction in (5.9) calculates the average of the payoffs in all conditions that is considered to be the contribution of

Table 5.1 The Shapley value of player 1

Order	$v(S \cup \{1\}) - v(S)$
Order 1,2,3	$v(\{1\}) - v(\emptyset)$=0-0=0
Order 1,3,2	$v(\{1\}) - v(\{\emptyset\})$=0-0=0
Order 2,1,3	$v(\{2,1\}) - v(\{1\})$=0-0=0
Order 3,1,2	$v(\{3,1\}) - v(\{1\})$=0-0=0
Order 2,3,1	$v(\{2,3,1\}) - v(\{2,3\})$=1-0=1
Order 3,2,1	$v(\{3,2,1\}) - v(\{3,2\})$=1-0=1

the player to the coalition. Note that the Shapley value has nothing to do with the cost of the players. For example, consider three players with costs of 10, 20, and 30 units, respectively, and their payoff function:

$$v(S) = \begin{cases} 1 & : \quad S = \{1,2,3\} \\ 0 & : \quad otherwise, \end{cases}$$

The number of orders in this example is 6. Different orders lead to different payoffs. Table 5.1 shows the Shapley value calculation for the 1st player which is $\frac{1}{3}$. In a similar way, we can compute the Shapely values of the 2nd and 3rd players. It can be seen that all the players' Shapely values are $\frac{1}{3}$ which implies that their contributions for the coalition are the same even though their costs are different.

From (5.9), if two micro grids have equal contributions to the coalition, their Shapley values are the same, although their individual values could be different. Also, the Shapely value does not depend on the order of the micro grid in the coalition. Even though the result may look unfair, it indicates the practical contribution of the micro grids to the coalition.

Using the adopted strategy for objective selection and the concept of Shapely value for extra profit allocation in a coalition, an algorithm can now be designed to formulate coalitions of the micro grids. First, we need to introduce a necessary definition from [22].

Definition. Assume two collections of disjoint coalitions $\mathcal{A} = \{A_1, \ldots, A_i\}$ and $\mathcal{B} = \{B_1, \ldots, B_j\}$ that are formed out of the same players. For one collection $\mathcal{A} = \{A_1, \ldots, A_i\}$, the payoff of a player k in a coalition $A_k \in \mathcal{A}$ is $\phi_k(\mathcal{A}) = \phi_k(A_k)$ where $\phi_k(A_k)$ is given for coalition A_k. Collection $\mathcal{A}$ is preferred over $\mathcal{B}$ by *Pareto order*, i.e. $\mathcal{A} \triangleright \mathcal{B}$, if and only if

$$\mathcal{A} \triangleright \mathcal{B} \Leftrightarrow \{\phi_j(\mathcal{A}) \geq \phi_j(\mathcal{B}), \forall k \in \mathcal{A}, \mathcal{B}\} \tag{5.10}$$

with at least one strict inequality ($>$) for a player k. The Pareto order implies that a group of players prefer to join a collection $\mathcal{A}$ rather than $\mathcal{B}$, if at least one player can improve her payoff when the structure is changed from $\mathcal{B}$ to $\mathcal{A}$ without curtailing the payoff of any other player.

To form the coalition, *merge* and *split* [24] rules are defined:

Definition Merge. Merge any set of coalitions $\{S_1, \ldots, S_l\}$ where $\{\cup_{i=1}^{l} S_i\} \rhd \{S_1, \ldots, S_l\}$, therefore, $\{S_1, \ldots, S_l\} \rightarrow \{\cup_{i=1}^{l} S_i\}$.

Definition Split. Split any coalition $\{\cup_{i=1}^{l} S_i\}$ where $\{\{S_1, \ldots, S_l\} \rhd \cup_{i=1}^{l} S_i\}$, therefore, $\{\cup_{i=1}^{l} S_i\} \rightarrow \{S_1, \ldots, S_l\}$.

From the above merge and split definitions, notice that some micro grids will join a new coalition or merge with a bigger coalition, respectively, given that at least one of them can improve payoff without cutting down the payoffs of any other micro grid or coalition, respectively. On the other hand, a large coalition could split into several smaller coalitions (or in worst case disappear) if the micro grids find that they can leave the coalition or merge with a smaller coalition so as to get more payoff than that the current coalition. So, a merge or split decision by Pareto order makes sure that all the involved micro grids agree on it.

By exploiting the merge and split operations, we now describe our algorithm in the following steps.

1. Each micro grid could obtain information of others (e.g., neighboring micro grids' position, buying/selling status, available power, etc.) by using the communication infrastructure or communication technology of smart grid (i.e., smart meters, BAN/NAN GW, control center, etc).
2. The micro grids will generate the energy, receive the user demands, and estimate whether to buy or sell energy.
3. The forming coalition stage begins. For a specific partition $\mathcal{S} = \{S_1, \ldots S_k\}$, each coalition $S_i \in \mathcal{S}$ communicates to its neighbors, uses merge and split operation rules to find the best cooperative partners to form a bigger coalition or leave the bigger coalition to form a smaller one so as to get more profits (improve payoff). The coalitions of micro grids calculate their payoffs by employing (5.7) and (5.8), find that the payoffs of all of them will increase, and this is the Pareto order in Eq. (6.20), if they can form a coalition.
4. After the merge and split iterations, the system will compose of disjoint coalitions, and no coalitions may have any further incentive to perform further merge or split operation.

Upon such convergence, the micro grids within each formed coalition will start their energy transfer.

5.4 Numerical Results

In this section, simulation results are presented to evaluate the effectiveness of our adopted game-theoretic coalition formulation approach for the micro grids. The considered scenario has a power distribution grid topology of $10 \times 10 \, \text{km}^2$. The macro station is at the center of the grid topology while the micro grids are

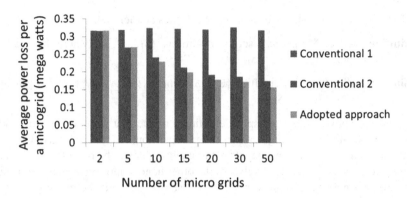

Fig. 5.2 Comparative result of the conventional methods and the adopted coalition-based power loss minimization approach

deployed randomly. Req_i is set to a random value varying from -20 to $20\,\mathrm{MW}$ (negative and positive signs mean whether MG_i is a buyer or seller, respectively). R is set to $0.2\,\Omega/\mathrm{km}$, $\alpha = 0.02$ is set to assumption in [25], and U_0 and U_1 are set to 50 and $22\,\mathrm{kV}$, respectively, to reflect practical values in existing smart grid distribution networks [25]. The price of the each of the unit power loss parameters is set to $w_1 = 1$ and $w_2 = 3$ (Fig. 5.2).

By varying the numbers of the micro grids from 2 to 50, the average power loss per micro grid is plotted in Fig. 7.3. Our adopted approach is compared with the non-cooperative case and also a conventional algorithm called NMS which are referred to as conventional 1 and conventional 2, respectively. When the number of micro grids increases, the average power loss changes just a little in the non-cooperative conventional 1 case. On the other hand, in case of conventional 2 and the adopted approach, the power losses decrease with the growing numbers of micro grids. When the number of micro grids is 50, the power loss in the adopted approach reaches up to significant reduction compared to conventional 2 approach. This superior performance happens because the power losses within the formed coalitions in the adopted approach are much lower than those between the macro station and the micro grids. As a result, when most of the micro grids have joined coalitions, the overall cost of the users decreases, and thus, the adopted approach outperforms conventional 2 in terms of improvement of the average power loss.

To reflect the money saving of the users, Fig. 5.3 demonstrates the optimal number of micro grids in different cases. Remember that the optimal number of micro grids depends on the area of the considered area and the demands of users in that area. Therefore, with increasing demands, the users require more electricity from the micro grids to reduce their cost (compared to procuring electricity from the macro station). For instance, the optimal number changes from 15 to 50 when the demands changes from 15 to $55\,\mathrm{MW}$ in the considered grid area of $10 \times 10\,\mathrm{km}^2$. Furthermore, because the resistance is a linear increasing function of distance, a higher resistance means a higher power loss. So, in a larger area, more micro grids

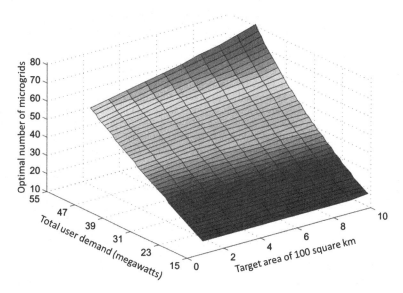

Fig. 5.3 The optimal number of MGs in the considered topology

want to minimize the higher power loss. Although not shown in the figure, as the area changes from 100 to $1000\,\text{km}^2$ and the demands are 15 MW, the optimal number of micro grids to form coalitions becomes from 15 to 17.

5.5 Concluding Remarks

In this chapter, we adopted a game-theoretic coalition formulation strategy for distributed micro grids. The strategy allows the micro grids to form coalitions so that the power loss is minimized when energy is transmitted from a micro grid to other micro grids or the macro station. The adopted approach also permits the micro grids to make decisions on whether to form or break the coalitions while maximizing their utility functions through alleviating the energy transfer power loss. Through simulation results, the effectiveness of the adopted approach was shown.

References

1. D. Niyato and P. Wang, "Cooperative transmission for meter data collection in smart grid," *IEEE Communications Magazine*, vol. 50, no. 4, pp. 90–97, Apr. 2012.
2. L. Ochoa and G. Harrison, "Minimizing energy losses: Optimal accommodation and smart operation of renewable distributed generation," *IEEE Transactions on Power Systems*, vol. 26, no. 1, pp. 198–205, Feb. 2011.

3. M. Erol-Kantarci, B. Kantarci, and H. Mouftah, "Cost-aware smart microgrid network design for a sustainable smart grid," in *2011 IEEE GLOBECOM Workshops (GC Wkshps)*, Dec. 2011, pp. 1178–1182.
4. S. Meliopoulos, G. Cokkinides, R. Huang, E. Farantatos, S. Choi, Y. Lee, and X. Yu, "Smart grid infrastructure for distribution systems and applications," in *2011 44th Hawaii International Conference on System Sciences (HICSS)*, Kauai, Hawaii, Jan. 2011, pp. 1–11.
5. S. Deilami, A. Masoum, P. Moses, and M. Masoum, "Real-time coordination of plug-in electric vehicle charging in smart grids to minimize power losses and improve voltage profile," *IEEE Transactions on Smart Grid*, vol. 2, no. 3, pp. 456–467, Sep. 2011.
6. A. Vargas and M. Samper, "Real-time monitoring and economic dispatch of smart distribution grids: High performance algorithms for DMS applications," *IEEE Transactions on Smart Grid*, vol. 3, no. 2, pp. 866–877, Jun. 2012.
7. A. Costabeber, T. Erseghe, P. Tenti, S. Tomasin, and P. Mattavelli, "Optimization of micro-grid operation by dynamic grid mapping and token ring control," in *Proceedings of the 2011-14th European Conference on Power Electronics and Applications (EPE 2011)*, Birmingham, UK, Aug. 2011, pp. 1–10.
8. W. Saad, Z. Han, and H. Poor, "Coalitional game theory for cooperative micro-grid distribution networks," in *2011 IEEE International Conference on Communications (ICC) Workshops*, Seattle, Washington, USA, Jun. 2011, pp. 1–5.
9. A. Costabeber, P. Tenti, and P. Mattavelli, "Surround control of distributed energy resources in micro-grids," in *2010 IEEE International Conference on Sustainable Energy Technologies (ICSET)*, Kandy, Sri Lanka, Dec. 2010, pp. 1–6.
10. P. Tenti, A. Costabeber, D. Trombetti, and P. Mattavelli, "Plug amp; play operation of distributed energy resources in micro-grids," in *32nd International Telecommunications Energy Conference (INTELEC)*, Orlando, Florida, USA, Jun. 2010, pp. 1–6.
11. G. Corso, M. Di Silvestre, M. Ippolito, E. Sanseverino, and G. Zizzo, "Multi-objective long term optimal dispatch of distributed energy resources in micro-grids," in *2010 45th International Universities Power Engineering Conference (UPEC)*, Cardiff, United Kingdom, Aug. 2010.
12. M. Kirthiga, S. Daniel, and S. Gurunathan, "A methodology for transforming an existing distribution network into a sustainable autonomous micro-grid," *IEEE Transactions on Sustainable Energy*, vol. 4, no. 1, pp. 31–41, Jan. 2013.
13. Z. Li, C. Wu, J. Chen, Y. Shi, J. Xiong, and A. Wang, "Power distribution network reconfiguration for bounded transient power loss," in *2012 IEEE Innovative Smart Grid Technologies - Asia (ISGT Asia)*, Tianjin, China, May 2012, pp. 1–5.
14. J. Aguero, "Improving the efficiency of power distribution systems through technical and non-technical losses reduction," in *Transmission and Distribution Conference and Exposition (T&D), 2012 IEEE PES*, Dallas, Texas, USA, May 2012, pp. 1–8.
15. L. McDonald, R. Storry, A. Kane, F. McNicol, G. Ault, I. Kockar, S. McArthur, E. Davidson, and M. Dolan, "Minimisation of distribution network real power losses using a smart grid active network management system," in *2010 45th International Universities Power Engineering Conference (UPEC)*, Cardiff, United Kingdom, Aug. 2010.
16. K. Nakayama and N. Shinomiya, "Distributed control based on tie-set graph theory for smart grid networks," in *2010 International Congress on Ultra Modern Telecommunications (ICUMT) and Control Systems and Workshops*, Moscow, Russia, Oct. 2010, pp. 957–964.
17. T.-I. Choi and Y.-K. Cho, "International business offering related to innovative smart technologies," in *2012 IEEE PES Innovative Smart Grid Technologies (ISGT)*, Washington D.C., USA, Jan. 2012, pp. 1–8.
18. M. Ismail, M.-Y. Chen, and X. Li, "Optimal planning for power distribution network with distributed generation in Zanzibar island," in *2011 International Conference on Electrical and Control Engineering (ICECE)*, Yichang, China, Sep. 2011, pp. 266–269.
19. N. Katie, V. Marijanovic, and I. Stefani, "Profitability of smart grid solution application in distribution network," in *7th Mediterranean Conference and Exhibition on Power Generation, Transmission, Distribution and Energy Conversion (MedPower 2010)*, Agia Napa, Cyprus, Nov. 2010, pp. 1–6.

20. Z. M. Fadlullah, Y. Nozaki, A. Takeuchi, and N. Kato, "A survey of game theoretic approaches in smart grid," in *2011 International Conference on Wireless Communications and Signal Processing (WCSP)*, Nanjing, Jiangsu, China, Nov. 2011, pp. 1–4.

21. C. Zhang, W. Wu, H. Huang, and H. Yu, "Fair energy resource allocation by minority game algorithm for smart buildings," in *Design, Automation Test in Europe Conference Exhibition (DATE), 2012*, Dresden, Germany, Mar. 2012, pp. 63–68.

22. H. Li and W. Zhang, "QoS routing in smart grid," in *2010 IEEE Global Communications Conference (GLOBECOM 2010)*, Miami, Florida, USA, Dec. 2010.

23. C. Wei, Z. M. Fadlullah, N. Kato, and A. Takeuchi, "GT-CFS: a game theoretic coalition formulation strategy for reducing power loss in micro grids," *IEEE Transactions on Parallel and Distributed Systems*, vol. 25, no. 9, pp. 2307–2317, Sep. 2014.

24. K. R. Apt and A. Witzel, "A generic approach to coalition formation," *Computing Research Repository - CoRR*, vol. abs/0709.0435, 2007. [Online]. Available: http://arxiv.org/abs/0709.0435

25. J. Machowski, J. W. Bialek, and J. R. Bumby, *Power system dynamics : stability and control.* Chichester, U.K. Wiley, 2008, rev. ed. of: Power system dynamics and stability / Jan Machowski, Janusz W. Bialek, James R. Bumby. 1997. [Online]. Available: http://opac.inria.fr/record=b1135564

Chapter 6
On Optimally Reducing Power Loss in Micro-Grids with Power Storage Devices

6.1 Background

Chapter 5 described a strategy to allow the micro grids to make coalitions and exchange power with other micro grids and/or the macro station. In this chapter, we are going to carry on from that strategy and enhance the model further by considering micro grids to use power storage devices. Recently, the micro grid developers and operators are exhibiting a great deal of interest in using lithium-ion batteries and flow batteries. The battery technologies have become quite mature and they are currently capable enough to provide exceptional renewable power integration in the micro grids based energy systems [1, 2]. In [3], it was demonstrated how the storage devices can supplement energy generation to consumption to achieve a balance between energy demand and supply within the micro grid. Necessity of optimal control of the power storage devices of the micro grid was also indicated in [3]. A new photovoltaic power generation and load power consumption prediction algorithm was designed in [4] which was specifically designed for a residential storage controller. However, these existing works usually focused on a single micro grid and did not consider the power losses impacting the entire system including the macro station and multiple micro grids. Instead, they had a localized approach such as how to reduce the power loss within a given micro grid, how to charge and discharge the storage device periodically, and so forth. In this chapter, we adopt a different approach by focusing on a scalable total power loss minimization approach across the entire smart grid.

6.2 System Model

Our considered power storage equipped micro grids model is shown in Fig. 6.1. As shown in the figure, at the center of the model, there is the macro station, which is the utility operator of a given region. The macro station could be one of the retailers

© The Author(s) 2015   53
Z.Md. Fadlullah, N. Kato, *Evolution of Smart Grids*, SpringerBriefs
in Electrical and Computer Engineering, DOI 10.1007/978-3-319-25391-6_6

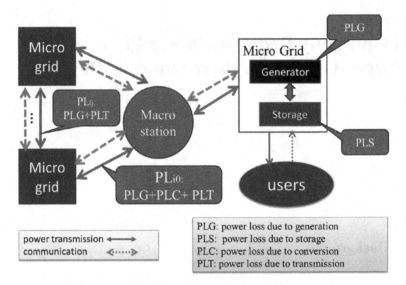

Fig. 6.1 Our considered micro grids system model and the various power losses occurring in the system

or even the power wholesaler. The macro station has its own customers such as residential users, factories, etc. Additionally, it is connected to the micro grids which are located in its area of operation. There are power distribution lines between the macro station and these distributed micro grids. Thus, power can be exchanged between the macro station and the micro grids. In normal mode of operation, the macro station sells power to the micro grids. However, if needed (particularly in peak hours), the macro station may also buy surplus power of the micro grids. The micro-grids are the secondary power generation sources in the considered area. They can use diesel and also integrate with simple or hybrid renewable energy sources like solar, wind power, small hydro, geothermal, waste-to-energy, hydraulic, and combined heat and power systems. The micro grids have their own customers (which may overlap with the customer base of the macro station) in their locality. The power produced in a micro grid is transmitted to its users according to their demand. The micro grids are assumed to have power storage devices like lithium-ion batteries, rechargeable car batteries, flywheel energy storage, etc. The storage devices get charged up during the off-peak time and discharge during the peak hours to meet the energy demand of the customers. In addition to their initial deployment cost, the power storage devices suffer from storage power loss during charging and discharging. Therefore, it is important to consider the storage power loss of the micro grid along with power generation and transmission losses.

Assume $\mathcal{N}$ is the set of micro-grids, and $N \triangleq |\mathcal{N}|$. Also assume that during t^{th} hour, the total user-demand of the ith micro grid MG_i ($i \in \mathcal{N}$) is $D_i(t)$, and MG_i can supply power $U_i(t)$ by discharging its charged power storage devices. Our model takes into account four types of power losses—generation power loss (see [5]), storage power loss [6], transmission power loss, and conversion power loss [7]. Even though other types of power losses are possible, they are usually negligible and our model does not include them. Assume for power m, the associated power losses are storage power loss $PLS(m)$, generation power loss $PLG(m)$, and transmission power loss $PLT(m)$. The general observation is that $PLS(m) \leq PLG(m) \leq PLT(m)$. Let the generation power be denoted by $G_i(t)$. Consider $G_i(t)$ and $D_i(t)$ as random numbers in the real smart grid networks from the work in [8]. To account for the generation power loss, the supply power $U_i(t)$ can be expressed as:

$$U_i(t) = (1 - \theta)G_i(t), \tag{6.1}$$

where θ is a fraction of the generation power loss. Based on (6.1), the generation power loss during t^{th} hour denoted by $PLG_i(t)$ is given by:

$$PLG_i(t) = \theta G_i(t) \tag{6.2}$$

The generation power of MG_i has a maximum and a minimum [8]. This means that its power generation $G_i(t)$ has a range.

$$G_{min} \leq G_i(t) \leq G_{max}, \tag{6.3}$$

where G_{min} and G_{max} are the minimum and maximum of the generated power by MG_i.

Many factors influence the quantity of power in the power storage devices like power self-discharge, the cost of power charging, etc. To quantify the storage power loss $PLS_i(t)$ to account for those factors, a function of charged power $S_i(t-1)$ in the time interval $(t - 1)$ is adopted:

$$PLS_i(t) = \beta S_i(t - 1), \tag{6.4}$$

where $S_i(t - 1)$ denotes the charged power of the power storage devices in time $(t - 1)$, and β is a fraction of the storage power loss. Also, the power storage device has a maximum capacity. To avoid over-discharge, the storage power $S_i(t)$ needs to satisfy the following condition:

$$S_{min} \leq S_i(t) \leq S_{max}, \tag{6.5}$$

where S_{min} and S_{max} are the minimum and maximum of the stored power, respectively. When the supply power is more than the users' power demand, some power will be charged in the power storage devices. Otherwise, the power storage devices will discharge to satisfy the demand. The storage power is calculated as:

$$S_i(t) = (1 - \beta)S_i(t - 1) + S_i^*(t), \tag{6.6}$$

where $S_i^*(t)$ denotes the operation of charging or discharging during t^{th} hour. $S_i^*(t)$ needs to satisfy the following.

$$S_i^*(t) = \begin{cases} \leq 0 & : \quad discharge \\ > 0 & : \quad charge, \end{cases} \tag{6.7}$$

When the storage power and supply power fall below the user-demand, or when the micro grid has surplus power (to sell), the micro grid needs to exchange power with the macro station and/or with its neighboring micro grids. Let $W_i(t)$ be the power which MG_i wants to exchange with other micro grids or the macro station:

$$W_i(t) = S_i^*(t) + U_i(t) - D_i(t), \tag{6.8}$$

where $U_i(t)$ and $S_i^*(t)$ are given by (6.1) and (6.7), respectively. If $(W_i < 0)$, the supply power and storage power of MG_i is below its users' aggregate demand. So, MG_i needs to buy power from other micro grids or the macro station. If $(W_i > 0)$, MG_i has surplus power which it can sell other micro grids in need of energy, or even to the macro station.

Now let us consider the other types of power losses. When power transmission happens between the macro station and micro grids, transmission power loss and conversion power loss happen [7]. In addition, generation power loss needs to be considered. Because to comply MG_i's request for additional energy, other micro grids or the macro station need to generate the power which will be associated with generation power loss. Thus, the total power loss between MG_i and the macro station can be expressed as:

$$PL_{i0}(t) = PLT_{i0}(t) + PLC_{i0}(t) + PLG_{i0}(t) \tag{6.9}$$

According to Chap. 5, $PLT_{i0}(t) = \frac{B_i(t)^2 R_{i0}}{U_0^2}$ and $PLC_{i0}(t) = \alpha B_i(t)$. Therefore, (6.9) can be re-written as:

$$PL_{i0}(t) = \frac{B_i(t)^2 R_{i0}}{U_0^2} + (\alpha + \theta)B_i(t), \tag{6.10}$$

where R_{i0} is the power line resistance between MG_i and the macro-station, U_0 denotes the transmission voltage between MG_i and the macro-station, α is a fraction of the conversion power loss, and $B_i(t)$ indicates the power that MG_i wants to exchange with the macro station. $B_i(t)$ could be any one of the following:

$$B_i(t) = \begin{cases} W_i(t) & : \quad W_i(t) > 0 \\ L_i^*(t) & : \quad W_i(t) < 0 \\ 0 & : \quad Req_i = 0, \end{cases} \tag{6.11}$$

where $L_i^*(t)$ is the total amount of power which needs to be generated to ensure that MG_i can obtain the power required to meet its demand W_i. L_i^* is the solution of following quadratic equation:

$$L_i(t) = PL_{i0}(t) + |W_i(t)| = \frac{R_{i0}L_i^2(t)}{U_0^2} + (\alpha + \theta)L_i(t) - W_i(t). \tag{6.12}$$

For a given $L_i^*(t)$, three possible solution sets of (6.12) exist (zero, one, and two solutions). To minimize the value of $L_i^*(t)$, we choose the smaller one when (6.12) has two roots. If (6.12) has no solution, assume that the root is the same as (6.12) with a single root, which is $L_i^*(t) = \frac{(1-\alpha-\theta)U_0^2}{2R_{i0}}$.

Similar with (6.10), the power loss between the micro grids can be estimated. Since the voltage among the micro grids are medium level voltage, the conversion power loss can be somewhat neglected. Therefore, if MG_i and MG_j want to exchange power, the power loss function PL_{ij} can be expressed as follows.

$$PL_{ij}(t) = PLT_{ij}(t) + PLG_{ij}(t) = \frac{R_{ij}B_{ij}^2(t)}{U_1^2} + \theta B_{ij}(t), \tag{6.13}$$

where R_{ij} is the resistance of the distribution line between MG_i and MG_j. U_1 is the transmission voltage between MG_i and MG_j. Then, B_{ij} can be expressed as:

$$B_{ij}(t) = \begin{cases} B_i(t) & : & |B_i(t)| \leq |B_j(t)| \\ B_j(t) & : & otherwise, \end{cases} \tag{6.14}$$

where B_i and B_j are given by (6.11). It implies that if the seller MG_i cannot meet the demand of the buyer MG_j, then MG_i only sells B_i to MG_j. Also, MG_j will buy at least $\frac{U_1^2}{2R_{ij}}$ amount of power from MG_i due to the power loss between MG_i and MG_j.

Finally, the total power losses $PTA_i(t)$ of MG_i can be given by:

$$PLA_i(t) = PLG_i(t) + PLS_i(t) + PL_{i0}(t) + \sum_j PL_{ij}(t) \tag{6.15}$$

Therefore, the total payoff function of the coalition C is as follows.

$$u(C, \Pi, t) = -\sum_i PLA_i(t) \tag{6.16}$$

where $\Pi \in C_C$ denotes the join order of the micro grids, which decide to join the coalition C, and C_C is the set of the micro-grids' order in C. If a micro grid wants to maximize its payoff, it has to minimize the total power losses. This means that the micro grid wants the maximum of (6.16). Hence, by using (6.16), which indicates the total power losses incurred by the different power transmissions for C, it is possible to define the value function for the micro grids $(\mathcal{N}, v)$ coalition game:

$$v(C, t) = \max_{\Pi \in \mathcal{C}_C} u(C, \Pi, t). \tag{6.17}$$

Consider the micro grids as "players" as part of a cooperative game. When the strategies are given, these players can adopt the best choice. The strategy of the players and the coalition game which they play are described next.

6.3   Coalition Formulation Strategy for Micro Grids with Power Storage Devices

Based on the earlier described system model, the micro grids could have different strategies or choices to make coalitions. Each micro grid, however, confronts the challenge of how to find the best strategy from a set of all possible strategies. For example, if a micro grid wants to make a coalition with other micro grids, it needs to know which micro grids are the most appropriate ones. A game-theoretic approach is adopted to overcome this issue [9].

The total power losses inside a micro grid comprise generation power loss (from renewable sources of the micro grid) and storage power loss, respectively. Consider this as the "inner power loss." On the other hand, the "external power loss" happens when power is exchanged among the micro grids or between a micro grid and the macro station. Hence, the external power loss comprises generation, transmission, and conversion power losses. From the system model, we understand that the inner power loss is less than the external power loss in the same condition. Additionally, the storage power loss is lower than the generation power loss inside the micro grids. Hence, when a micro grid's storage power, its aggregate user-demand, and the generated power are given, the minimum of the inner power loss, the operation of charging and discharging S_i^*, and the required power W_i can be estimated. The next challenge is how to form the coalition to optimally reduce the total power losses.

If MG_i wants to join the coalition C, it will increase the payoff of the coalition. The "marginal payoff function" per unit power of MG_i for the coalition is expressed as:

$$MPF(i, C) = \frac{u(C + \{i\}, \Pi^*, t) - u(C, \Pi, t)}{B_{iC}}, \tag{6.18}$$

where $C + \{i\}$ denotes the new set including set C and MG_i. Π^* and Π are the join orders for the new and old sets, respectively. B_{iC} is the power which MG_i wants to exchange with the coalition C. This means extra profits per unit power when MG_i joins the coalition C. Because different micro grids will lead to different MPFs, based on (7.10), the micro grids can make the best decisions to form coalition.

When the coalitions are formulated, because power loss among the micro grids is less than that between the macro station and the micro grids, there are extra payoffs. In order to distribute these extra payoffs, the "Shapley value" [10] is adopted (similar to its use in Chap. 5). When there is a coalition game $(\mathcal{N}, v)$, the Shapley value of player i (i.e., MG_i) can be calculated using:

$$\phi_i(v) = \sum_{C \subseteq \mathcal{N} \setminus \{i\}} \frac{|C|!(N - |C| - 1)}{N!} (v(C \cup \{i\}) - v(C)), \qquad (6.19)$$

where N is the number of players and the sum extends over all the subsets C of $\mathcal{N}$ without the ith player. $v(C)$ is given by (6.17). For the micro grids, it is the contribution of forming the coalition. Therefore, the extra payoffs will be distributed based on (6.19).

The set of micro grids can be divided into two disjoint subsets (set of buyers C_b and set of sellers C_s) depending on the value of W_i. C_b consists of the micro grids MG_p ($p \in |C_b|$) with ($W_p < 0$), and C_s is composed of the micro grids MG_q ($q \in |C_s|$) with ($W_q > 0$). If ($W_i = 0$), then MG_i can be considered as either a buyer or a seller. To maximize the payoffs of coalitions, each micro grid finds its appropriate partners so as to maximize (6.17) using the following steps.

- Initialization: At first, $S_i(t), D_i(t)$ and $G_i(t)$ are given for the tth hour. The operation $S_i^*(t)$ and the demand $W_i(t)$ are evaluated. The set $\mathcal{N}$ is divided into C_s and C_b. C_s and C_b are sorted in descending order according to the requests of the selling or buying micro grids, i.e., $C_s = \{s_1, \ldots, s_k\}$. The sum of sets is calculated and the lesser one of the two sets is determined.

  For ease of explanation, assume that C_s is the lesser one. Then, select $s_l \in C_s$ as the objective.
- Step 1: Based on (7.10), find the *appropriate* micro grids in C_s or C_b to form coalition C with an objective that the profit of coalition C is the maximum. This step indicates that the power loss of micro grids in coalition C is less than that between the macro station and the micro grids belonging to C.
- Step 2: If the remainder of C_s is less than that of C_b, choose the biggest one in C_s. Go to step 1 until there is no availability in the sets or one of the sets becomes empty.
- Step 3: If the remainder of C_s is more than that of C_b, choose the biggest one in C_b. Go to step 1, until there is no availability in the sets or one of the sets becomes empty.
- Step 4: Calculate the "Shapley value" of the micro grids, and distribute the extra payoffs.

Based upon the aforementioned steps for objective selection and the concept of Shapely value for extra profit distribution in a coalition, we adopt an algorithm to formulate distributed coalitions of micro grids in the remainder of this section. Before the algorithm can be formally described, let us introduce an important definition from [8].

Definition. Consider two collections of disjoint coalitions $\mathcal{A} = \{A_1, \ldots, A_i\}$ and $\mathcal{B} = \{B_1, \ldots, B_j\}$ that are formed out of the same players. For one collection $\mathcal{A} = \{A_1, \ldots, A_i\}$, the payoff of a player k in a coalition $A_k \in \mathcal{A}$ is $\eta_k(\mathcal{A}) = \eta_k(A_k)$ where

$\eta_k(A_k)$ is given by (6.17) for coalition A_k. Collection $\mathcal{A}$ is preferred to $\mathcal{B}$ by *Pareto order*, i.e. $\mathcal{A} \triangleright \mathcal{B}$, if and only if

$$\mathcal{A} \triangleright \mathcal{B} \Leftrightarrow \{\eta_k(\mathcal{A}) \geq \eta_k(\mathcal{B}), \forall k \in \mathcal{A}, \mathcal{B}\}, \qquad (6.20)$$

with at least one strict inequality ($>$) for a player k.

The Pareto order signifies that a group of players prefer to join a collection $\mathcal{A}$ instead of $\mathcal{B}$, if at least one player is able to improve its payoff when the structure is changed from $\mathcal{B}$ to $\mathcal{A}$ without reducing the payoff of any other.

To form the coalition, two distributed rules are needed: *merge* and *split* [11] defined as follows:

Definition Merge. Merge any set of coalitions $\{S_1, \ldots, S_l\}$ where $\{\cup_{i=1}^l S_i\} \triangleright \{S_1, \ldots, S_l\}$, hence, $\{S_1, \ldots, S_l\} \rightarrow \{\cup_{i=1}^l S_i\}$.

Definition Split. Split any coalition $\{\cup_{i=1}^l S_i\}$ where $\{\{S_1, \ldots, S_l\} \triangleright \cup_{i=1}^l S_i\}$, hence, $\{\cup_{i=1}^l S_i\} \rightarrow \{S_1, \ldots, S_l\}$.

The merge and split definitions help micro grids to maximize their payoffs, and find the proper micro grids to form coalitions. Since the micro grids act as players of a cooperative game, we adopt a coalition formation algorithmic approach for micro grids with power storage devices by exploiting the merge and split operations. First, in our adopted approach, each micro grid is able to share information (e.g., its position, neighboring micro-grids, etc.) with the others using the communication infrastructure or communication technology. Second, the micro grids generate the power, get the demands of the users, charge/discharge power into/from their storage devices, and take a decision whether to buy or sell the power. Third, the coalition formulation stage begins in which the merge or split processes occur as follows. Given a partition $\mathcal{C} = \{C_1, \ldots C_k\}$, each micro grid in each coalition $C_i \in \mathcal{C}$ communicates with its neighboring micro grids. Using these negotiations, the coalitions exchange the information with others that include the numbers of coalitions, their capabilities, etc. The information helps the coalitions to find the best cooperative partners to make a larger coalition or leave the larger coalition to create some smaller ones to obtain more profits (payoffs). The coalitions or micro-grids calculate their payoffs using (6.16) and (6.17). When they find that all their payoffs will increase by merging into a particular coalition, they reach the Pareto order as shown in Eq. (6.20).

6.3.1   Characterization of the Game

Characterization of the adopted game involves proving its stability, convergence, and optimality.

Definition. A coalition C: = $\{C_1, \ldots, C_k\}$ is $\mathbb{D}_{hp}$-stable if the following two conditions are satisfied.

(a) for each $i \in \{1, \ldots, k\}$ and for each partition $\{P_1, \ldots, P_l\}$ of the coalition C_i:
$v(C_i) \geq \sum_{j=1}^{l} v(P_j)$.
(b) for each set $T \subseteq \{1, \ldots, k\}$: $\sum_{i \in T} v(C_i) \geq v(\cup_{i \in T} C_i)$ [12].

Lemma 6.3.1. *The coalition formed by the proposed algorithm is $\mathbb{D}_{hp}$-stable [7].*

Lemma 6.3.2. *The considered $(\mathcal{N}, v)$ micro grids coalition game converges to the Pareto optimal $\mathbb{D}_{hp}$-stable partition, if such a partition exists. Otherwise, the final partition is merge-and-split proof [7].*

Lemma 6.3.3. *The total power loss incurred in the proposed scheme is minimum.*

Proof (Reduction to Absurdity). Assume that the total power loss is not minimized. If the adopted approach can help the micro grids to obtain demand W and corresponding total power loss PLA is not minimum, there exists a demand W^* which is different from W and the corresponding total power loss PLA^* is minimum. Then, $PLA < PLA^*$, if $W > W^*$. It is because that the inner power loss is less than the external power loss under the same condition. It means that the use of storage power or generation power is better than purchased power from other micro grid(s)/the macro station. Therefore, PLA^* is not minimum. If $W < W^*$, it means that there is power surplus while the demand is not satisfied. This condition is not reasonable since it arises contradiction. Hence, the total power loss in the adopted approach is minimum.

Theorem 6.3.4. *There exists a global optimal result for each micro grid using the adopted approach.*

Proof. From Lemmas 6.3.1 to 6.3.3, we know that the model results in a global optimal result.

From Theorem 6.3.4, the Pareto optimal is the only stable situation. So, the merge and split operations help the micro grids to maximize their utilities (i.e., minimize the total power loss) until the $\mathbb{D}_{hp}$-stable situation occurs. In such a situation, no micro grid can decrease its total power losses without increasing the total power losses of the other micro grids.

6.4 Numerical Results

Some comparative results are included to show the effectiveness of our adopted approach compared with a conventional non-cooperative method [4]. In the conventional method, the micro grids only exchange power with the macro station and they cannot exchange power with the other micro grids. Our considered scenario considers a power distribution grid topology, area of which is $10 \times 10 \, \text{km}^2$. The macro station is placed at the center of the grid while the micro grids are arbitrarily deployed in the topology. The resistance between the micro grids is the same as that between the macro station and any micro grid, and its value is set to $R = 0.2\,\Omega/\text{km}$. The fraction of power transmission α, the fraction of power generation θ, and the fraction of power storage β are set to 0.02, 0.05, 0.01, respectively based on [5, 6, 13]. The power demand D_i of MG_i is derived from a uniform distribution between 10 and 742 MW based on [8]. The power generation G_i is obtained using a uniform distribution between 10 and 326 MW. Suppose that the capacity of power storage device is 200 MW, and the minimum storage power is 10 MW. U_0 and U_1 are set to 50 and 22 kV, respectively, based on [13]. The prices of a unit power loss are set as $w_1 = 1$ and $w_2 = 3$ [7]. In the adopted approach, the users send the information to the corresponding micro grids, and the micro grids also exchange the information to other micro grids or the macro station if necessary. Assume that the micro grids can communicate with the macro station through an optical backbone network, capacity of which is 100 Mbps. For simplicity, each micro grid is assigned to satisfy the demand of 100 users. The packet size for user to micro grid communication is set to 102 bytes [14] while that for inter-micro grid communication is considered to be 112 bytes.

Figure 6.2 shows the average power loss per micro grid for different numbers of the micro-grids ranging from 5 to 50. In the conventional method, the power loss per micro grid does not improve since the micro grids only obtain power from the macro station. On the other hand, in case of the adopted approach, the average power loss is significantly improved with the increasing numbers of the micro grids. This happens because the coalitions are formed by the micro grids with the objective of optimally alleviating the power loss. When the micro-grids could successfully make the coalitions in the adopted approach, they could exchange power with other micro grids instead of the macro station leading to the reduction of the average power loss.

Figure 6.3 demonstrates that the micro grids want to purchase the power from the macro station for $N = 20$ micro grids in the conventional method as well as in the adopted approach. Suppose that the peak period of the day is from 12 p.m. to 9 p.m.

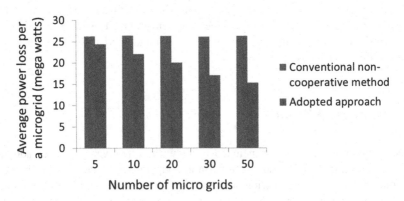

Fig. 6.2 Comparative average power loss per micro grid achieved in the conventional method and the adopted approach

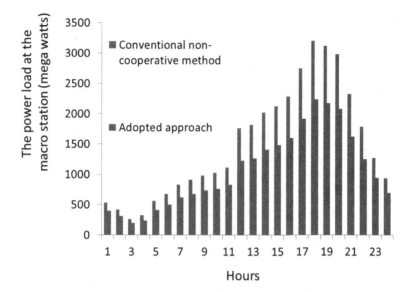

Fig. 6.3 Comparison of the power load on the macro station in the conventional method and the adopted approach

Also suppose that the situations of the micro grids are remain fixed since their initial random deployment in the grid. Even though in both schemes, the micro grids have the power storage devices that they could charge in off-peak time and discharge in peak time, notice that compared with the conventional method, the result achieved (i.e., the burden in terms of the power load inflicted upon the macro station) by the adopted game-theoretic approach is lower. This happens because the micro grids in the adopted approach can buy power from their neighboring micro grids instead of the distant macro station. Thus, both the users and the macro station can obtain benefits from forming coalitions through the adopted approach.

6.5 Concluding Remarks

In this chapter, a game-theoretic coalition formulation method for micro grids with power storage devices was discussed. Our adopted approach allows the micro grids to take a decision on whether to charge or discharge their power storage devices, and to find other micro grids (i.e., appropriate neighbors) to build coalitions to optimally minimize the total power losses. The adopted approach also allows the micro grids to make decisions on whether to form or split the coalitions while alleviating the power losses due to power generation, transmission, and storage. In addition, it was shown that the adopted approach offers a stable, convergent, and optimal solution. Furthermore, the effectiveness of the adopted approach was demonstrated compared to a conventional method.

References

1. S. Minnihan, "Microgrids focus on security, creating needs for UPS storage," accessed Aug. 2015. [Online]. Available: http://www.smartgridnews.com/story/microgrids-focus-security-creating-needs-ups-storage/2013-05-29
2. C. Hill, M. Such, D. Chen, J. Gonzalez, and W. Grady, "Battery energy storage for enabling integration of distributed solar power generation," *IEEE Transactions on Smart Grid*, vol. 3, no. 2, pp. 850–857, Jun. 2012.
3. C. Ahn and H. Peng, "Decentralized voltage control to minimize distribution power loss of microgrids," *IEEE Transactions on Smart Grid*, vol. 4, no. 3, pp. 1297–1304, Sep. 2013.
4. Y. Wang, X. Lin, and M. Pedram, "Accurate component model based optimal control for energy storage systems in households with photovoltaic modules," in *2013 IEEEGreen Technologies Conference*, Denver, CO, USA, Apr. 2013, pp. 28–34.
5. E. Bendict, T. Collins, D. Gotham, S. Hoffman, and D. Karipdes, "Losses in Electric Power Systems," accessed Aug. 2015. [Online]. Available: http://docs.lib.purdue.edu/cgi/viewcontent.cgi?article=1270&context=ecetr
6. IEA-ETSAP and IRENA Technology Policy Brief E18, "Electricity storage," accessed Aug. 2015. [Online]. Available: https://www.irena.org/DocumentDownloads/Publications/IRENA-ETSAP%20Tech%20Brief%20E18%20Electricity-Storage.pdf
7. C. Wei, Z. M. Fadlullah, N. Kato, and A. Takeuchi, "GT-CFS: a game theoretic coalition formulation strategy for reducing power loss in micro grids," *IEEE Transactions on Parallel and Distributed Systems*, vol. 25, no. 9, pp. 2307–2317, Sep. 2014.
8. Z. Li, C. Wu, J. Chen, Y. Shi, J. Xiong, and A. Wang, "Power distribution network reconfiguration for bounded transient power loss," in *2012 IEEE Innovative Smart Grid Technologies - Asia (ISGT Asia)*, Tianjin, China, May 2012, pp. 1–5.
9. C. Wei, Z. M. Fadlullah, N. Kato, and I. Stojmenovic, "On optimally reducing power loss in micro-grids with power storage devices," *IEEE Journal on Selected Areas in Communications*, vol. 32, no. 7, pp. 1361–1370, Jul. 2014.
10. L. S. Shapley, "A value for n-person games," *Contributions to the theory of games*, vol. 2, pp. 307–317, 1953.
11. K. R. Apt and A. Witzel, "A generic approach to coalition formation," *Computing Research Repository - CoRR*, vol. abs/0709.0435, 2007. [Online]. Available: http://arxiv.org/abs/0709.0435
12. K. R. Apt and T. Radzik, "Stable partitions in coalitional games," *Computing Research Repository - CoRR*, vol. abs/cs/0605132, 2006. [Online]. Available: http://arxiv.org/abs/cs/0605132
13. J. Machowski, J. W. Bialek, and J. R. Bumby, *Power system dynamics : stability and control*. Chichester, U.K. Wiley, 2008, rev. ed. of: Power system dynamics and stability / Jan Machowski, Janusz W. Bialek, James R. Bumby. 1997. [Online]. Available: http://opac.inria.fr/record=b1135564
14. M. M. Fouda, Z. M. Fadlullah, N. Kato, R. Lu, and X. Shen, "A lightweight message authentication scheme for smart grid communications," *Smart Grid, IEEE Transactions on*, vol. 2, no. 4, pp. 675–685, Dec. 2011.

Chapter 7
A Distributed Paradigm for Power Loss Reduction in Micro-Grids

7.1 Background

In the earlier two chapters, we showed that in order to enhance the efficiency of the smart grid in terms of power loss reduction, the micro grids can exchange power with other micro grids instead of the macro station. If the micro grids have more than one neighbor, which neighbor can be chosen as the most appropriate partner for building the power loss reducing coalition is a key challenge [1, 2]. The coalition formation approaches in the earlier chapters are actually subject to central decision making. In other words, the micro grids have to take their problem up to the control center of the macro station which evaluates all the parameters and then makes optimal decisions pertaining to micro grids' coalition formation. Then, the control center has to send the decision back to the micro grids. This means that the micro grids have to go back and forth with the control center since it is the only entity which has the global information (e.g., which micro grids have energy shortage and by how much, which micro grids have surplus power and how much they can sell in the next hour, etc.). In this chapter, we discuss a distributed decision making approach to suit the physically distributed micro grids to allow them to make autonomous decisions regarding energy transfer and coalition formation. The distributed decision making may not be optimal as the centralized version. But it may save precious time for the micro grids to take prompt decisions rather than relying on the centralized entity to make decisions on their behalf. The adopted approach is essentially different from other existing decentralized algorithms like [3, 4] in terms of its objective to minimize the power losses.

© The Author(s) 2015
Z.Md. Fadlullah, N. Kato, *Evolution of Smart Grids*, SpringerBriefs
in Electrical and Computer Engineering, DOI 10.1007/978-3-319-25391-6_7

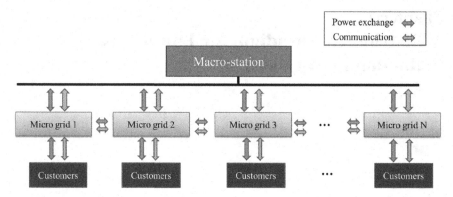

Fig. 7.1 The considered system model

7.2   System Model and Problem Formulation

Our proposed system model is described with the aid of Fig. 7.1. Consider that
there are three layers in this model. The top layer consists of the macro station,
which is typically the regional utility operator. The macro station can also be the
energy wholesaler or the regional retailers. A single macro station is shown in the
figure, but multiple macro stations are also possible on the top layer. The macro
station is able to exchange power with the secondary power sources on the second
layer. These secondary power sources are the micro grids with renewable/hybrid
power generation sources. For simplicity, suppose that the macro station has enough
power to satisfy the demands of the micro grids and receive the surplus power
from the micro grids. Following Chap. 6, each micro grid is physically linked to
the macro station. However, compared with the macro station, the micro grids could
be deployed much nearer to the customers, who are in the third and bottom layer of
the considered model. Therefore, the customers could be linked to the micro-grids
directly. The micro grids supply power to the customers to satisfy their demands.
Additionally, the micro grids can exchange power with their neighbors or even with
the macro-station, when supply and demand are unequal. Because they just know
the location of their neighbors, a distributed algorithm could be adopted. A purely
distributed algorithm could help the micro grids to find proper partners so as to
minimize the total power loss of the smart grid. The smart meters are installed at
the customer-premise through which the customers can send the demands to the
micro-grids.

Let $\mathcal{N}$ indicate the set of the micro-grids and $N = |\mathcal{N}|$. In the given time period
(e.g., one second), for the ith micro grid denoted by MG_i, we define a real function
$D_i(t)$ as the current remaining power of MG_i and it can be expressed as:

$$D_i(t) = G_i(t) - W_i(t). \tag{7.1}$$

Here, $G_i(t)$ and $W_i(t)$ are the generation power and the aggregate user-demand of MG_i, respectively. It means that MG_i wants to obtain power to satisfy its demand ($D_i(t) < 0$). MG_i has a power surplus to sell ($D_i(t) > 0$), or its supply equals its demand ($D_i(t) = 0$). The micro grids can be divided into two types, namely "exporters" and "importers" (synonymous to sellers and buyers, respectively, as described in Chaps. 5 and 6). The exporters have surplus to sell while the importers need additional amount of power to satisfy the customer-demands. If the current remaining power of MG_i is zero, MG_i is considered to be either an exporter or an importer, and it cannot have any impact. Additionally, the demand of customers $W_i(t)$ and production power $G_i(t)$ are always considered as random numbers in the real smart grid networks [5]. Hence, the value of $D_i(t)$ is considered as a random number with a certain observed distribution.

When $D_i(t) \neq 0$, MG_i will exchange power with other micro grids or the macro station. It will then lead to power loss. For simplicity, two types of power losses are considered in our model even though other types of losses can be accommodated as mentioned in Chap. 6. The first one is power loss due to transmission (PLT) [5]. To alleviate the power loss of long-distance transmission, the voltage between the primary power station to the macro station is high (e.g., 50 kV [5]) while that between the micro grid and the macro station is medium (e.g., 22 kV [5]). Therefore, high voltage needs to be converted into medium level. This process will cause the power loss due to conversion (PLC).

First, the power loss between two micro-grids are considered. Based on [5], if MG_i transmits power to MG_j, the power loss function $P_{ij}(t)$ can be expressed as:

$$P_{ij}(t) = \frac{R_{ij}Q_{ij}^2(t)}{U_1^2}, \tag{7.2}$$

where R_{ij} is the resistance of the distribution line between MG_i and MG_j. U_1 is the transfer voltage between MG_i and MG_j, and it is less than U_0. In this model, we do not consider the power loss of transforming voltage levels between MG_i and MG_j. Also, $Q_{ij}(t)$ is defined as:

$$Q_{ij}(t) = \begin{cases} \frac{Q_{ij}^2(t)R_{ij}}{U_1^2} - D_j(t) : |D_i(t)| > |D_j(t)| \\ D_i(t) \qquad\qquad :otherwise. \end{cases} \tag{7.3}$$

If MG_i sells power to MG_j, the current remaining power $D_i(t)$ will be updated as:

$$D_i(t) = D_i(t) - Q_{ij}(t). \tag{7.4}$$

If MG_i buys power from MG_j, $D_i(t)$ is updated as follows:

$$D_i(t) = \min\{D_i(t) + Q_{ij}(t) - PL_{ij}(t), 0\}. \tag{7.5}$$

After exchanging power with the other micro grids, if $D_i \neq 0$, MG_i will exchange power with the macro station. In this process, consider PLT and PLC ([5]). If MG_i wants to sell $D_i(t)$ to the macro station ($D_i(t) > 0$) or buy $D_i(t)$ from the macro-station ($D_i(t) < 0$), we are able to express the power loss $PL_{0i}(t)$ as follows:

$$PL_{0i}(t) = \frac{R_{0i}Q_{0i}^2(t)}{U_0^2} + \alpha Q_{0i}(t), \tag{7.6}$$

where R_{0i} is the distribution line resistance between the macro-station and MG_i, U_0 is the voltage of power transfer between MG_i and the macro station, and α is a fraction of power loss caused by voltage conversion. For simplicity, α is treated as a constant [5]. $Q_{0i}(t)$ is the power that MG_i wants to buy or sell.

The value of $Q_{0i}(t)$ is any of the following.

$$Q_{0i}(t) = \begin{cases} \frac{Q_{0i}^2(t)R_{ij}}{U_0^2} + \alpha Q_{0i}(t) - D_i(t) & : \quad D_i(t) < 0 \\ D_i(t) & : \quad otherwise. \end{cases} \tag{7.7}$$

Based on (7.2) and (7.6), in a given time-slot t, the total power loss $PLA_i(t)$ of MG_i is given by:

$$PLA_i(t) = PL_{0i}(t) + \sum_j \frac{PL_{ij}(t)}{2}. \tag{7.8}$$

If MG_i exchanges power with MG_j, the power loss $PL_{ij}(t)$ should not be calculated twice. Therefore, $PLA_i(t)$ includes half of $PL_{ij}(t)$.

Our target in this chapter is to allow the micro grids to make distributed decisions so as to minimize the total power loss. Hence, the objective function is

$$Minimize \sum_i PLA_i(t)$$

$$s.t. \quad D_i(t) \leq G_i(t) + \eta_i(t) \quad \forall i \in \mathcal{N}, \tag{7.9}$$

where $\eta_i(t) = sign(D_i(t))Q_{0i}(t) - PL_{0i}(t) + \sum_j(sign(D_i(t))Q_{ij}(t) - PL_{ij}(t))$, $sign(D_i(t)) = 1$ if $D_i(t) < 0$, and $sign(D_i(t)) = -1$ otherwise. Hence, our condition is that the demand at each micro grid does not exceed the sum of the amount of the remaining produced power and the power it exchanged with other micro grids and the macro station.

7.3   The Distributed Solution Approach for the Micro Grids

Based on the system model in Sect. 7.2, the total power loss of the smart grid could be calculated. However, unlike the centralized algorithms considered in Chaps. 5 and 6, the micro grids do not need to acquire the total information (i.e., the global snapshot of all other micro grids and all the parameters). Assume that the micro grids only have knowledge of the locations of one-hop neighbor(s), and are able to exchange power with it/them.

At the beginning of time-slot t, MG_i estimates $W_i(t)$ from the aggregate demand of its customers. To satisfy $W_i(t)$, MG_i generates power $G_i(t)$. If the current remaining power of MG_i is $D_i(t) \neq 0$, MG_i will exchange power with its neighbor(s). MG_i will exchange information of $D_i(t)$ with its neighbors. Based on the remaining power of neighbors, MG_i generates a set of Potential Exchange power Neighbors (PEN). This set implies that if $MG_j \in PEN$, MG_i has the opportunity to exchange power with MG_j. If PEN of MG_i has more than one element, it needs to select a proper neighbor to exchange power so as to minimize the total power loss. The "Reducing power loss per Unit exchanged Power" (RUP) of MG_i and MG_j for the micro grid pair can deal with this problem. If MG_i exchanges power with MG_j, the function can be expressed as:

$$RUP(Q_{ij}(t)) = \frac{PL_{0i}(t) + PL_{0j}(t) - PL_{ij}(t)}{|Q_{ij}(t)|}. \tag{7.10}$$

The above function represents potential extra payoffs (reducing power loss) per unit exchange power, if MG_i joins the coalition. $PL_{0i}(t)$ and $PL_{0j}(t)$ indicate power losses if the same power $Q_{ij}(t)$ was exchanged with the macro station by both micro grids in the current coalition. Merging them could replace these two by power exchange between them, with power loss $PL_{ij}(t)$. Higher values of RUP imply saving power per unit power. Therefore, based on (7.10), the micro grids can make the best decisions to merge their coalitions.

MG_i calculates RUP (Eq. (7.10)) of PEN and sorts PEN in a descending order according to RUP. Then, MG_i considers the first element j from PEN, if PEN is not empty. MG_i sends $D_i(t)$ to MG_j, waits for the response from MG_j unless time-out occurs. If MG_i receives "accept" response from MG_j, it will exchange power with MG_i, based on $D_i(t)$ and $D_j(t)$, delete j from PEN, and update $D_i(t)$. MG_j will be deleted when MG_i is waiting for the response from MG_j and time-out occurs, or the response is "reject." At the same time, MG_i receives offers from its neighbors as well. If MG_i is waiting for the response from the first element of PEN j and receives the offer from other neighbor $k(k \neq j)$, the status of MG_k will be set as "hold," and the hold message is returned. The neighbor k will not be deleted until time-out occurs. The above action will not repeated until $D_i(t) = 0$ or PEN is empty. After exchanging power with one-hop neighbors, $D_i(t)$ is updated by the quantities of exchanged powers. If $D_i(t) \neq 0$, MG_i will exchange power with the macro station.

Algorithm 1 Distributed algorithm for micro grids' power exchange. Input: $W_i(t)$, $G_i(t)$, Output: $Q_{ij}(t)$, $Q_{0i}(t)$, and $PLA_i(t)$

1: BEGIN
2: for each MG_i
3: Loop
4:     Estimate $D_i(t)$, based on $W_i(t)$, $G_i(t)$ (Eq. 7.1)
5:     While $(D_i(t) \neq 0)$
6:     Share $D_i(t)$ with neighbors of MG_i.
7:     Calculate PEN, RUP of PEN, and sort PEN
8:     While $(D_i(t) \neq 0$ and PEN is not empty$)$
9:        Get first element of PEN j
10:       If (j.status=hold and time-out occurs)
11:          Remove j from PEN
12:       Else
13:       Send offer to MG_j and wait for response
14:       while (time-out does not occur)
15:          If (the response from j ="accept")
16:            exchange power with j, calculate $Q_{ij}(t)$, $PL_{ij}(t)$ based on Eqs. 7.2, 7.3, update $D_i(t)$
             and remove j from PEN.
17:          Endif
18:          If (receive offer from neighbor k and $k \neq j$)
19:            k.status=hold and send "hold" message to k
20:          Endif
21:       Endwhile
22:       Delete j from PEN
23:       Endif
24:     Endwhile
25:     If $(D_i(t) \neq 0)$
26:        Compute $Q_{0i}(t)$, $PL_{0i}(t)$ based on 7.6 and 7.7 and $D_i(t)$=0
27:     Else
28:        If received offer from neighbor l
29:          Send "reject" to neighbor l
30:        Endif
31:     Endif
32:     Endwhile
33: Endloop
34: END

These steps are summarized in Algorithm 1 [6]. For example, suppose that there are one macro station and five micro grids (denoted by MG1 to MG5 as shown in Fig. 7.2). In this example, $RUP_{12} = 1.2$, $RUP_{23} = 2$, $RUP_{34} = 3.5$, and $RUP_{45} = 4$. Based on these RUPs, MG1 will send offer to MG2, MG2 will send offer to MG3, MG3 will send offer to MG4, and MG5 will send offer to MG4. Because MG4 sends offer to MG5 and waits for the response, MG4 will send "hold" to MG3. In the same way, MG3 and MG2 send "hold" to MG2 and MG1, respectively. When MG4 receives response from MG5, they will exchange power. Because $PL_{45} = 0.2$, $Q_{45} = 1.3$. After that $D_4 = D_5 = 0$ and MG4 sends "reject" to MG3. When MG3 receives "reject," it will activate the offer of MG2 and exchange power with MG2

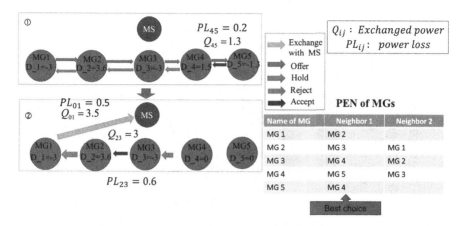

Fig. 7.2 A simple example showing how the distributed algorithm 1 enables power exchange between micro grids and macro station with minimized power loss

$(Q_{23} = 3)$. Hence, MG1 receives "reject" from MG2, after power transmission between MG2 and MG3 $(D_2 = D_3 = 0)$. Eventually, MG1 will exchange power with the macro station $(Q_{01} = 3.5)$.

Theorem 7.3.1. *The solution of Algorithm 1 is Pareto Optimal.*

Proof. Assume that the solution $(a_1, a_2, \ldots, a_N)$ is not Pareto Optimal. Therefore, there exists at least a micro grid $l \in N$, which can adjust its action a_l to a_l^* so as to augment its utility while utilities of others will not be diminished. In other words, $u(a_l, a_{-l}) < u(a_l^*, a_{-l})$. Because the algorithm could help micro grids to find the most appropriate neighbors and exchange power to maximize their payoff, the micro grids cannot augment their utilities by changing the solution of the algorithm. Therefore, $u(a_l, a_{-l}) \geq u(a_l^*, a_{-l})$ $\forall a_l^* \in A_l$. This result contradicts the previous assumption that the solution is not Pareto Optimal.

7.4 Numerical Results

In this section, numerical results are demonstrated to show the effectiveness of the adopted distributed algorithm compared to (1) the distributed algorithm in [5] whereby the micro grids choose the nearest neighboring micro-grid to exchange power with, and (2) the centralized approach in Chap. 6. Our considered scenario comprises a power distribution grid topology, area of which $10 \times 10 \, \text{km}^2$. The macro station is deployed at the center of the grid, and the micro grids are deployed randomly in the topology. Each micro grid is linked with its one hop neighboring micro grid and the macro station. Similar to the assumption made by Saad et al. [5], the power demands of the customers $W_i(t)$ of MG_i is derived from a Gaussian

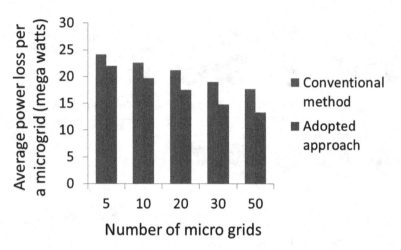

Fig. 7.3 Comparison of the average power loss in the conventional method and the adopted distributed algorithmic approach

distribution between 10 and 316 MW. The power generation $G_i(t)$ is obtained from a Gaussian distribution between 10 and 316 MW. The resistance between the micro grids is the same as that between the macro station and any micro grid. R is set to $0.2\,\Omega$/km. The fraction of power conversion parameter α is set to 0.02 based on [7]. U_0 and U_1 are set to 50 and 22 kV, respectively based on the practical values considered in [7]. The prices of the each of the unit power are set as $w_1 = 1$ and $w_2 = 3$ [1].

Figure 7.3 demonstrates the average power loss per micro grid for different numbers of micro grids ranging from 5 to 50 in case of the conventional method and our adopted distributed algorithmic approach. When the number of micro grids increases, the power losses decrease in both schemes. However, the distributed approach shows better performance because all the relevant power loss components are considered to achieve near-optimal minimization of the total power losses while finding the micro grids their suitable one-hop neighbors to exchange power.

Figure 7.4 shows the percentage of cost saving using the adopted distributed approach in contrast with the conventional method. When the number of micro grids increases, the percentage becomes larger. This is because the adopted technique helps the micro grids to discover proper neighbors so as to minimize the total power losses, and thereby, saves money. Thus, it allows the entire power grid to save a significant amount of money in contrast with the conventional method.

Figure 7.5 demonstrates that the macro station in the conventional method needs to supply more power for the micro girds to satisfy their user-demands than that in the adopted distributed approach. The reason is that in the conventional case, the micro grids only exchange power with the nearest one-hop neighboring micro grids without considering the total power loss while by using the adopted distributed algorithm, the micro grids can exchange power with others so as to reduce the total

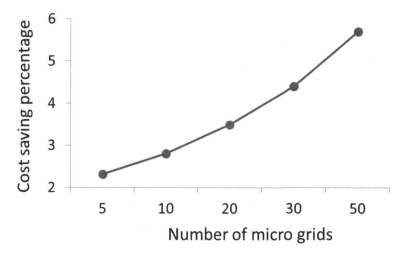

Fig. 7.4 The percentage of cost saving using the adopted distributed approach compared with the conventional method

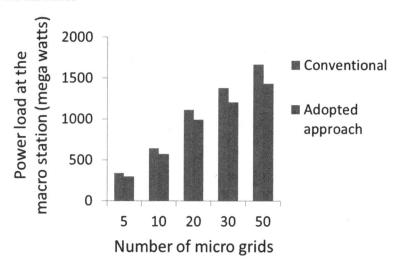

Fig. 7.5 Comparison of the power load on the macro station in the conventional case and the adopted distributed approach

power loss. Higher power loss will cause higher power load from the macro station. Therefore, the adopted approach helps the macro station to decrease the peak of power generation and improve its efficiency.

Furthermore, consider the comparison of the performance of the centralized method in Chap. 6 and the adopted distributed approach. In the centralized case, the micro grids send demands to the control center of the macro station. The macro station will help all the micro grids to find proper neighbors as it knows all the information of the micro grids. However, in the distributed approach, the micro grids

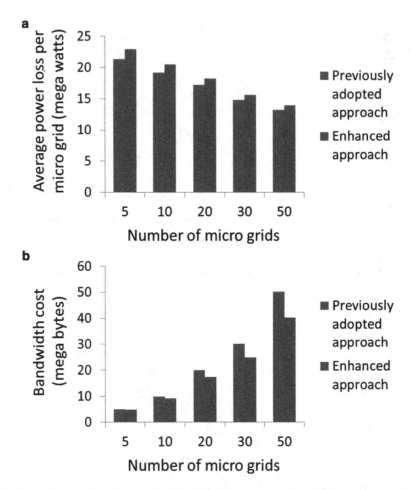

Fig. 7.6 The comparison in centralized and distributed approaches. (**a**) Average power loss. (**b**) The bandwidth cost

only know the demands of one hop neighbors. By using the distributed approach, the micro grids will send the demands to one-hop neighbors so as to minimize the total power loss. If time-out occurs, the micro grids will give up this neighbor and try to exchange power with other neighbors. Admittedly, this process could add to the total power loss. Therefore, the average power loss in the distributed approach is slightly higher than that in the centralized approach as shown in Fig. 7.6a. On the other hand, because the micro grids do not send information to the control center, the communication bandwidth cost in the distributed approach is lower than that in the centralized approach as demonstrated in Fig. 7.6b. It means that by using the distributed approach, precious communication bandwidth could be saved for more important tasks such as accommodating AMI-oriented communication of many more end-users in the micro grids.

7.5 Concluding Remarks

In this chapter, we discussed a novel cooperative power exchange algorithm for micro grids from a purely distributed perspective. This allows the micro grids to form coalitions so as to minimize the total power loss. Through numerical results, the effectiveness of the approach was verified. Comparative results showed its superior performance compared to a conventional distributed method. Also, some trade-off in the performance between the centralized approach and this distributed one could be seen.

References

1. C. Wei, Z. M. Fadlullah, N. Kato, and A. Takeuchi, "GT-CFS: a game theoretic coalition formulation strategy for reducing power loss in micro grids," *IEEE Transactions on Parallel and Distributed Systems*, vol. 25, no. 9, pp. 2307–2317, Sep. 2014.
2. C. Wei, Z. M. Fadlullah, N. Kato, and I. Stojmenovic, "On optimally reducing power loss in micro-grids with power storage devices," *IEEE Journal on Selected Areas in Communications*, vol. 32, no. 7, pp. 1361–1370, Jul. 2014.
3. M. E. Elkhatib, R. E. Shatshat, and M. M. A. Salama, "Decentralized reactive power control for advanced distribution automation systems," *IEEE Transactions on Smart Grid*, vol. 3, no. 3, pp. 1482–1490, Sep. 2012.
4. H. Liang, B. J. Choi, W. Zhuang, and X. Shen, "Stability enhancement of decentralized inverter control through wireless communications in microgrids," *IEEE Transactions on Smart Grid*, vol. 4, no. 1, pp. 321–331, Mar. 2013.
5. W. Saad, Z. Han, and H. Poor, "Coalitional game theory for cooperative micro-grid distribution networks," in *2011 IEEE International Conference on Communications (ICC) Workshops*, Seattle, Washington, USA, Jun. 2011, pp. 1–5.
6. C. Wei, Z. M. Fadlullah, N. Kato, and I. Stojmenovic, "A novel distributed algorithm for power loss minimizing in smart grid," in *2014 IEEE International Conference on Smart Grid Communications (SmartGridComm)*, Venice, Italy, Nov. 2014, pp. 290–295.
7. J. Machowski, J. W. Bialek, and J. R. Bumby, *Power system dynamics : stability and control.* Chichester, U.K. Wiley, 2008, rev. ed. of: Power system dynamics and stability / Jan Machowski, Janusz W. Bialek, James R. Bumby. 1997. [Online]. Available: http://opac.inria.fr/record= b1135564

Chapter 8
Security Challenge in the Smart Grid

8.1 Security Problem: A Bi-product of Legacy Communication Networks

Whether we consider the overall smart grid or the micro grids paradigm (as discussed in Chap. 2), security is an important issue which cannot be overlooked. The available smart metering technologies including the AMI often arise privacy concerns since they rely on centralizing personal energy consumption information of the users at their smart meters. In the Netherlands, there was a legal ruling in 2009 to make it mandatory to consider privacy issues of using smart meters [1]. Also, in the US policy adopted by NIST stated that there should be privacy for design approach for smart grid communications [2]. These privacy concerns can be addressed by appropriately authenticating the smart meters. Such a solution, however, needs to take into consideration the limited resources (like low memory and computational capacity) of the smart meters. So, any authentication scheme for smart grid communication requires careful design so as to enforce adequate security while placing minimal burden on the already limited resources of the smart meters. In this chapter, we discuss a light-weight message authentication method for securing communication among various smart meters at different points of the smart grid. Our adopted method is based on the Diffie–Hellman key establishment protocol and hash-based message authentication code that allows smart meters to make mutual authentication and achieve message authentication in a light-weight fashion. That is, it does not result in high delay and exchanges few signal messages in the message authentication phase.

Securing smart grid communication relies on two important factors [3]—communication delay and large volume of messages exchanged in the smart grid. For example, consider the case that the control center accidentally misses inputs from some HAN GWs. This may influence the control center's decision in upcoming energy distribution scheduling. To avoid any potential emergency situation which could occur at any time, the smart grid communication system needs to have

© The Author(s) 2015
Z.Md. Fadlullah, N. Kato, *Evolution of Smart Grids*, SpringerBriefs
in Electrical and Computer Engineering, DOI 10.1007/978-3-319-25391-6_8

Table 8.1 Example of power
requirements of different
loads in a typical HAN

Load	Power requirement (kW/h)
Air conditioner	1
Fridge	0.2
Microwave oven	0.1
Lighting arrays	0.05
Desktop computer	0.2

the ability to handle the message delivery to the control center via the BAN and
NAN GWs with the minimum possible delay. The power requirements of the HAN
devices/loads which are listed in Table 8.1 are sent to the respective BAN by meter
periodic data read (i.e., ToM#2 (revisit Fig. 2.3)). Each raw periodic request message
has a size of 32 bytes. If we consider the mandatory headers, the packet size can,
therefore, be roughly $(50 + 32 =)$ 82 bytes. Additionally, we need to consider
TCP/IP headers and optional security headers if any security protocol is used. If the
BAN GW experiences congestion, the packet may be delayed to be sent to the NAN
GW and control center. Furthermore, it may also get dropped if the RAM and the
on-chip flash of the BAN GW are full. The BAN GW memory can be full due to
several reasons like multiple messages arriving from different HANs at the same
time, limited processing capability of the BAN GWs means the buffer remains full
until the processing is complete, etc. If that is the case, then the BAN GW may
ask the HAN GW to retransmit the required packets. This could lead to increased
communication delay. Practically speaking, the smart grid communication delay
should be in the order of a few milliseconds according to [3, 4]. But it is tough to
achieve this in a large scale smart grid. So, how to minimize the communication
delay becomes one key challenge.

Hauser et al. [3] further report that the smart grid communication network needs
to be able to accommodate more messages simultaneously without having any major
impact on the communication delay. The large volume of messages in the smart grid
communication is likely to influence the available bandwidth. Consider a model
where a control center, connected with 10,000 feeders (and BAN GWs), serves
100,000 customers. Assume that each HAN GW generates a message every second
to the BAN GW [5] during peak hours. Then the total number of messages generated
per second is 100,000. Additionally, the BAN GWs generate messages to each other
and also to the control center through the NAN GW. If the average packet size is
100 bytes, the required transmission line bandwidth in the radio access network is
estimated to be 800 Mbps.

As demonstrated in the above example, any secure smart grid communication
framework requires to have light-weight operations to (1) avoid possibly high
communication delay, and (2) reduce communication overhead by cutting down
unnecessary signal messages. Also, remember that the security headers contribute
to the increased packet size as well (as illustrated in Fig. 2.3). This shows the impor-
tance of a light-weight authentication mechanism for HAN/BAN/NAN GWs in the
context of delay-sensitive and bandwidth-intensive smart grid communication.

However, the existing methods for facilitating smart grid communications security are often inherited directly from the other communication networks which do not necessarily take into account the context of specific communication requirement of the smart grid. Also, most security systems do not consider secure framework to reliably authenticate the smart meters operating on the different levels in the smart grid. For example, the BAN GW should authenticate the requesting HAN GWs while the NAN GW should be able to authenticate its BAN GWs. The cryptographic overheads may add a substantial portion to the packet size. Cryptographic operations also contribute to significant computation cost, particularly in the receiver-end which needs to verify the message. Thus, in the earlier example comprising 100,000 users, the number of messages to be verified per second by the NAN GW becomes significantly high. Also, there is processing delay at the respective smart meters to decrypt the arriving encrypted messages. This increases the communication delay. Because the conventional Public Key Infrastructure (PKI) methods are not adequate for the strict delay requirement of smart grid communications, a light-weight verification algorithm tailored for smart grid communications is needed so as to allow faster processing of incoming messages.

In addition, the smart meters are susceptible to various attacks found in literature. The use of IP-communication technologies in smart grid means that the power system can become vulnerable to cyber-attacks listed in Table 8.2. To mitigate this problem, a security framework is required, which can take into consideration various design objectives [6–10].

8.2 Light-Weight Authentication Mechanism

To address the earlier listed threats, we adopt a framework with security and reliability guarantees. The secure and reliable framework for smart grid communications have to achieve the following objectives.

1. Source authentication and message integrity: The smart meters need to be able to check the origin of a received packet and deliver the packet without any alteration. For instance, if a BAN GW is to receive a packet from a HAN GW, the BAN GW needs to authenticate the HAN GW. After successful authentication, it needs to check whether the packet is unaltered or not.
2. Small communication overhead and fast verification: The security scheme needs to be efficient in terms of low communication overhead and acceptable processing delay. In other words, a large number of message signatures from numerous smart meters need to be verified in a short time.
3. Conditional privacy preservation: The actual identity of a smart meter (which includes the owner's name, apartment number, etc.) needs to be concealed by appropriate encryption technology.

Table 8.2 Security threats against smart grid communications and the security requirements to address these problems

Attack	Impact on SG	Security requirement
Sniffing on smart meters	Same problem as conventional network	Encrypted packets: tougher for decoding traffic
Traffic analysis	Difficult to detect	Change encryption keys periodically
Denial of service (DoS), wireless jamming and interference	Can extract keys from second generation Zigbee chips [11]	Authenticated sharing of resources and/or channels
DoS buffer overflow attack	May delete the content of smart meters	Debug programs and protocol thoroughly
Reconfigure attack	Install unstable firmware on smart meter(s) and electronic appliances	Only permit secure firmware upgrade from authenticated CCs
Spoofing	– Impersonate smart meters – Increase victim's bill – Lower attacker's own bill	Authenticate smart meter over Internet Protocol Security (IPSec)
Man-in-the-Middle (MitM)	May impersonate smart meters [12]	Secure communication over IPSec
Replay attack	– Store current data (during low power usage) of smart meter – Then send the stored data to the utility company at a later time (during high power usage) [13]	– Use time-stamp and time-synchronization at smart meters and CC – Use time-variant nonce

4. Prevention of internal attack: A HAN GW owner should hold only her own cryptographic key, and she should not be able to access any neighboring HAN GWs' keying materials. Thus, even if a smart meter is compromised, an adversary cannot use the compromised smart meter to access other smart meters' sensitive information.
5. Maintaining forward secrecy: It should be assured that a session key derived from a set of long-term public and private keys will not be compromised if one of the (long-term) private keys is compromised in future.

Figure 8.1 shows our considered security framework for establishing a secure communication environment in the smart grid [8]. The framework has three parts—authentication, communication management, and network analysis, monitoring and protection. The smart meters are required to be authenticated before they can participate in the communication with other BAN/NAN GWs. The authentication scheme can be based on protocols such as Diffie–Hellman, SIGn-and-MAc

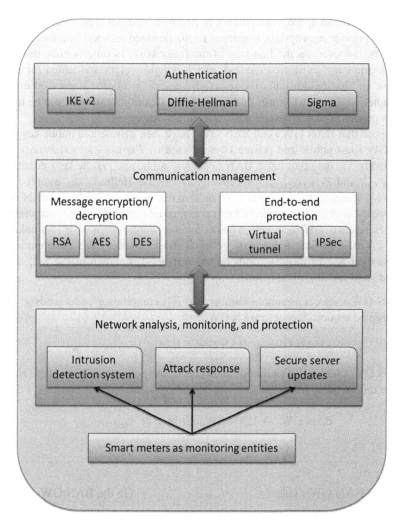

Fig. 8.1 Considered secure and reliable framework for smart grid communications

(SIGMA), and Internet Key Exchange (IKEv2). The communication management module has two parts—message encryption/decryption and end-to-end protection. Existing cryptographic algorithms like Data Encryption Standard (DES), Advanced Encryption Standard (AES), Rivest, Shamir, and Adleman (RSA) public key encryption, etc., can be used to encrypt the communication. For the end-to-end protection, Internet Protocol Security (IPSec) or virtual tunnel could be used. In the network analysis, monitoring, and protection module, smart meters are supposed to act as monitoring stubs which are equipped with anomaly and/or signature-based intrusion detection algorithms in order to detect malicious threats listed in Table 8.2. If the system finds any attack and detects a security update, it contacts a secure server to download appropriate patches or firmware updates. The monitoring stubs may also provide appropriate responses to the detected attacks. The focus of this chapter is, however, on the first step of the framework. In other words, designing an appropriate authentication method, which is light-weight and suited for delay-sensitive and bandwidth-intensive smart grid communications, is stressed upon in the chapter while other features of the smart grid security framework can be treated as open problems which require more attention in future.

Suppose that HAN GW i and BAN GW j have their private and public key pairs. HAN GW i has public and private keys denoted by Pub_{HAN_GWi} and $Priv_{HAN\_GWi}$, respectively. On the other hand, BAN GW j has public and private keys denoted by Pub_{BAN_GWj} and $Priv_{BAN\_GWj}$, respectively. The Diffie–Hellman key establishment protocol [14] is adopted for the initial handshake between the HAN and BAN GWs.

Suppose $\mathbb{G} = \langle g \rangle$ is a group of large prime order q such that the Computational Diffie–Hellman (CDH) assumption holds. In other words, given g^a, g^b, for unknown $a, b \in \mathbb{Z}_q^*$, it is difficult to estimate $g^{ab} \in \mathbb{G}$. Based on the CDH assumption, our adopted light-weight message authentication method is shown in Fig. 8.2, and its detailed steps are described below.

1. HAN GW i selects a random number $a \in \mathbb{Z}_q^*$, computes g^a, and sends g^a in an encrypted request packet to BAN GW j.

$$\text{HAN GW}_i \rightarrow \text{BAN GW}_j : \{i||j||g^a\}_{\{encr\}Pub_{BAN\_GWj}}$$

2. BAN GW j decrypts it and sends an encrypted response consisting of g^b. Here, b is a random number.

$$\text{BAN GW}_j \rightarrow \text{HAN GW}_i : \{i||j||g^a||g^b\}_{\{encr\}Pub_{HAN\_GWi}}$$

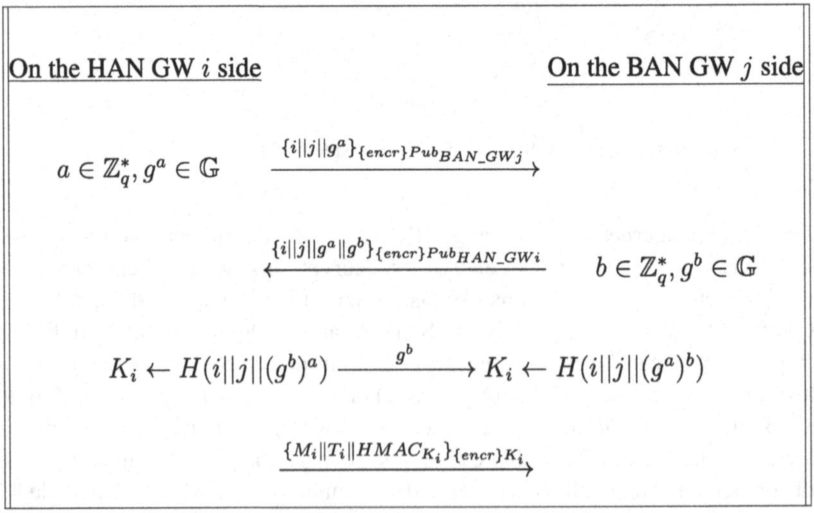

Fig. 8.2 Adopted light-weight message authentication scheme

3. After receiving BAN GW j's response packet, HAN GW i recovers g^a, g^b with its private key. If the recovered g^a is correct, BAN GW j is authenticated by HAN GW i. Then, with g^b and a, HAN GW i can calculate the shared session key $K_i = H(i||j||(g^b)^a)$, where $H : \{0, 1\}^* \rightarrow \mathbb{Z}_q^*$ is a secure cryptographic hash function, and sends g^b to BAN GW j in the plaintext form.
4. When the correct g^b is received by the BAN GW j, BAN GW j authenticates HAN GW i, and calculates the same shared session key $K_i = H(i||j||(g^a)^b)$.
5. To ensure data integrity in the late transmission, a Hash-based Message Authentication Code (MAC) generation algorithm using the shared session key K_i is utilized. The generated MAC denoted by $HMAC_{K_i}$ is based on the message M_i and recorded time instance of sending the message T_i, where T_i is used to prevent possible replay attacks. Then, HAN GW i transmits the following to the BAN GW j:

$$\text{HAN GW}_i \rightarrow \text{BAN GW}_j : \{M_i||T_i||HMAC_{K_i}\}_{\{encr\}K_i}$$

Since K_i is shared between BAN GW j and HAN GW i itself, BAN GW j can verify the authenticity of the sender and integrity of M_i. By this way, it can provide the NAN GW with the authenticated messages.

8.3 Security Analysis

Now let us analyze the security of the adopted light-weight message authentication method to verify whether the required security properties are satisfied.

- *The adopted method can provide mutual authentication.* In the adopted method, because g^a is encrypted with BAN GW j's public key, only if the adopted public key encryption technique is secure, then BAN GW j is the only entity who can recover g^a with the corresponding private key. So, when HAN GW i receives the correct g^a in Step 3, HAN GW i can realize that its counterpart is indeed BAN GW j. With the same reasoning, because g^b is encrypted with HAN GW i's public key, BAN GW j can also authenticate HAN GW i if it can receive the correct g^b in Step 4. Thus, the adopted method can provide mutual authentication between HAN GW i and BAN GW j.
- *The adopted method can establish a semantic-secure shared key in the mutual authentication environment.* The semantic security of the shared key under the chosen-plaintext attack means that an adversary $\mathcal{A}$ cannot distinguish the actual shared key K_i from the ones randomly drawn from the session key space, when $\mathcal{A}$ is given g^a, g^b and $Z \in \mathbb{G}$, where Z is either the actual shared key K_i or a random value R drawn from the session key space, according to a random bit $\beta \in \{0, 1\}$, i.e., $Z = K_i$ when $\beta = 0$, and $Z = R$ is returned when $\beta = 1$. Let $\beta' \in \{0, 1\}$ be $\mathcal{A}$'s guess on β. Then, the semantic security indicates $\Pr[\beta = \beta'] = \frac{1}{2}$. Now assume that there exists an adversary $\mathcal{A}$ that can break the semantic security of

the shared key with a non-negligible advantage $\varepsilon = 2\Pr[\beta = \beta'] - 1$ within the polynomial time. We can use the adversary $\mathcal{A}$'s capability to solve the CDH problem, i.e., give (g, g^a, g^b) for unknown $a, b \in \mathbb{Z}_q^*$, to estimate $g^{ab} \in \mathbb{G}$.

First, the adversary $\mathcal{A}$ is given (g, g^a, g^b), and also permitted to make q_H distinct queries on the random oracle $\mathcal{H}$ in the random oracle model [15]. To account for these random oracle queries, let us maintain a $\mathcal{H}$-list. When a new query $C_i \in \mathbb{G}$ is asked for the session key shared between HAN GW i and BAN GW j, we choose a fresh random number $Z_i \in \mathbb{G}$, set $\mathcal{H}(i||j||C_i) = Z_i$, put $(i||j||C_i, Z_i)$ in $\mathcal{H}$-list, and return Z_i to $\mathcal{A}$. When the adversary $\mathcal{A}$ makes a query on the session key, we flip a coin $\beta \in \{0, 1\}$, and return a random value $Z^* \in \mathbb{G}$.

Let $\mathcal{E}$ be the event that $C = g^{ab}$ has been queried by $\mathcal{A}$ to the random oracle $\mathcal{H}$. If the event $\mathcal{E}$ does not take place, $\mathcal{A}$ has no idea on the session key $K_i = H(i||j||g^{ab})$, then we have:

$$\Pr[\beta = \beta'|\bar{\mathcal{E}}] = \frac{1}{2}$$

and

$$\Pr[\beta = \beta'] = \Pr[\beta = \beta'|\mathcal{E}] \cdot \Pr[\mathcal{E}] + \Pr[\beta = \beta'|\bar{\mathcal{E}}] \cdot \Pr[\bar{\mathcal{E}}]$$

$$= \Pr[\beta = \beta'|\mathcal{E}] \cdot \Pr[\mathcal{E}] + \frac{1}{2} \cdot \Pr[\bar{\mathcal{E}}]$$

$$\leq \Pr[\mathcal{E}] + \frac{1}{2} \cdot (1 - \Pr[\mathcal{E}]) = \frac{1}{2} + \frac{\Pr[\mathcal{E}]}{2}$$

Additionally, because

$$\varepsilon = 2\Pr[\beta = \beta'] - 1 \Rightarrow \Pr[\beta = \beta'] = \frac{1}{2} + \frac{\varepsilon}{2}$$

we have $\Pr[\mathcal{E}] \geq \varepsilon$. Since $\mathcal{H}$-list contains q_H entries, we may pick up the correct $C_i = g^{ab}$ and solve the CDH problem with the success probability $1/q_H$ given that the event $\mathcal{E}$ does occur. Combining the above probabilities, we have:

$$\mathbf{Succ}^{\text{CDH}} = 1/q_H \cdot \Pr[\mathcal{E}] \geq \frac{\varepsilon}{q_H}.$$

However, this result is in contradiction with the CDH assumption. Hence, the adopted method can also establish a semantic-secure shared key. It is worth noting that if either HAN GW i or BAN GW j is compromised, the mutual authentication environment cannot be achieved. However, the compromise of either HAN GW i or BAN GW j's private key will not have impact on the security of the previous session keys. Therefore, the adopted method can also achieve perfect forward secrecy [14].

• *The adopted method can provide an authenticated and encrypted channel for the late successive transmission.* Since both HAN GW i and BAN GW j hold their shared session key K_i, the late transmission $\{M_i \| T_i \| HMAC_{K_i}\}_{\{encr\}K_i}$ can achieve integrity in addition to confidentiality. Additionally, the embedded timestamp T_i prevents possible replay attacks. Thus, the adopted method can provide an authenticated and encrypted channel for the late successive transmissions.

In summary, our adopted method is secure and suitable for the two-party communication in a smart grid environment.

8.4 Numerical Results

The effectiveness of the adopted message authentication method is evaluated using MATLAB [16]. For the smart grid topology in this evaluation, 10 NANs, each having 50 BANs, are considered. The number of HANs in each BAN is varied from a range of 10–140. Table 8.3 lists the other relevant parameters. The performance of our adopted method is compared with conventional ECDSA, which was used as a secure authentication protocol for smart grid demand response communications in [17]. In simulations, AES-128 algorithm was used to encrypt the packets to be transmitted using the shared session key, K_i, generated during the proposed authentication mechanism. In order to compare with this, we used ECDSA-256 authentication and encryption in simulations because its security level is comparable to that of 128-bits cryptography [18]. Note that only the messages exchanged between HANs and their corresponding BAN are considered for authentication. Furthermore, the session key is considered to be generated at the beginning of every fresh session.

The size of the HAN packet bound for the BAN is considered to be 102 bytes, which is sufficient to contain the users' power requirements and request to the control center. The size of the generated MAC is set to 16 bytes based on the RACE Integrity Primitives Evaluation Message Digest (RIPEMD-128)

Table 8.3 Simulation parameters used for evaluation.

Parameter	Value
BAN GW CPU clock	160 MHz
Number of HANs	10–140
HAN message generation interval (δ)	10 s
TCP header	20 bytes
Message header	50 bytes
Raw message	32 bytes
Hash header in proposed authentication scheme	16 bytes
ECDSA certificate size	125 bytes
ECDSA signature size	64 bytes
Simulation time	800 s

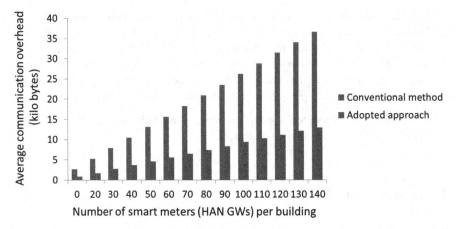

Fig. 8.3 Average communication overhead per BAN GW for varying number of smart meters (HAN GWs)

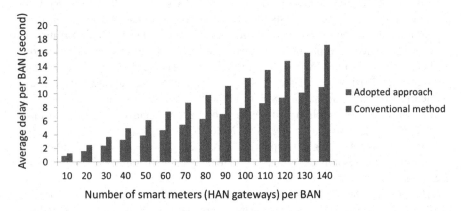

Fig. 8.4 Average delay at the BAN GW for varying number of smart meters (i.e., HAN GWs)

algorithm. This is done to utilize its resiliency against collision and pre-image attacks. The HAN message generation interval (δ) is considered to be 10 s to correspond with highly frequent need for demand-response communications in the smart grid. Communication overhead and message decryption/verification delay are compared in Figs. 8.3 and 8.4 for the conventional method and the adopted approach. Figure 8.3 shows the average communication overhead (in kilo bytes) experienced by the BAN GWs for varying numbers of smart meters. It should be noted that only one session per HAN GW with the BAN GW is supposed. When the number of smart meters is low, both the adopted approach and conventional method result in small overheads (below 5 kilo bytes). But the communication overheads start to gradually rise with the increase in smart meters. In case of the conventional method, the rise in the average communication overhead is much more significant than the adopted approach. The conventional method suffers from higher communication overheads mainly because of the certificate and signature included in each packet.

Next, Fig. 8.4 demonstrates the comparison between the adopted approach and the conventional method in terms of decryption/verification delay per BAN GW. OpenSSL package is used to measure the delays for both the adopted and conventional schemes [19] on a computer running 3.0 GHz Intel Xeon Processor (E5450) and Linux distribution of Debian 4.0. To simulate the BAN GW, the experimental values (e.g., decryption time) were scaled by 19.2 times to fit the 160 MHz of the BAN GW. As shown by the plot in Fig. 8.4, the decryption delay increases linearly for both the schemes. However, the conventional method shows higher decryption delay compared to that of the adopted approach since the latter provides a secure authentication process followed by AES encryption, which is faster than the conventional one relying on heavier signature verification plus decryption at the BAN for every message arriving from each HAN.

The memory usage of the adopted and conventional schemes over time for changing message volumes received by a given BAN GW is demonstrated in Fig. 8.5. The memory usage consists of two upper bounds, namely the RAM boundary and the RAM plus flash memory boundary that comprise 128 KB and 1 MB, respectively. Different message rates are used. The legends in the graph use the format of px and cx which mean the adopted approach and the conventional method with the message rate of x per δ. When the message rate is 50 per δ, the conventional method consumes about 50 kilo bytes of memory of the BAN GW. The adopted approach shows similar memory use. But notice when the number of

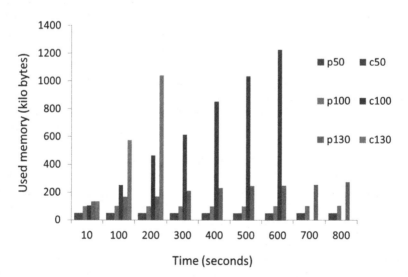

Fig. 8.5 Memory usage of the proposed and conventional ECDSA authentication algorithms for different message volumes received by BAN

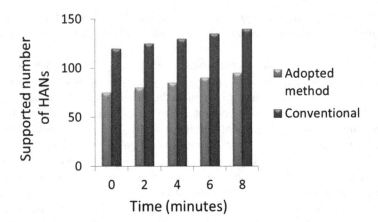

Fig. 8.6 Number of HANs supported by the proposed and conventional authentication schemes for smart grid communications

messages per δ arriving at the BAN GW increases to 100, the conventional method gets overwhelmed with the high number of messages coming from the high number of HANs, and it exceeds the RAM and flash memory bound after 570 s. Compared to this, the adopted approach uses much lower memory usage (approximately 100 kilo bytes) and continues to sustain this until end of simulation. However, when the number of apartments in a given building is further increased resulting in a higher message reception rate of 130 messages per δ at the BAN GW, the results change dramatically. In this case, the conventional method uses up all the available memory at the BAN GW very fast (within just 220 s of the beginning of the simulation). On the other hand, the adopted approach remains below 270 kilo bytes of the overall available memory throughout the conducted simulation. This superior performance of the adopted approach is because of its less processing in decrypting the packets that result in less queuing time in the RAM and the flash memory.

It may also be interesting to evaluate how may HANs are supported by the two compared schemes in terms of the available RAM and flash memory at the BAN GW over time. Figure 8.6 demonstrates this. In the conventional method, when the number of HANs per BAN exceeds 81, the memory usage starts to increase with time. This implies that after a while the memory usage will overflow the memory space of the BAN GW (i.e., 1152 kilo bytes consisting of 1 mega byte flash memory and 128 kilo bytes of RAM). At that point, the messages arriving from the HANs are dropped and not served within the BAN GW queue. For example, for 95 HANs supported by a particular BAN, the conventional method requires around 1260 kilo bytes of memory space to avoid any drop of messages. On the other hand, the adopted approach can accommodate 127 HANs within the BAN which is significantly higher than that achieved by the conventional method. This is possible because the adopted approach can process the messages arriving from the HANs in the BAN memory space much quicker than the conventional method.

8.5 Concluding Remarks

In this chapter, we described a light-weight message authentication scheme tailored for the requirements of smart grid communications based on Diffie–Hellman key establishment protocol and hash-based authentication code. A detailed security analysis of the adopted approach verifies its ability to satisfy the desirable security requirements within a secure and reliable smart grid communications framework. Also, comparative results demonstrate its superior performance over a conventional method.

References

1. C. Cuijpers and B.-J. Koops, "Het wetsvoorstel 'slimme meters': een privacytoets op basis van art. 8 evrm." Tilburg University, the Netherlands, Tech. Rep., Oct. 2008, accessed Aug. 2015. [Online]. Available: http://www.consumentenbond.nl/morello-bestanden/pdf-algemeen-2008/onderzoek_UvT_slimme_energi1.pdf
2. US National Institute for Standards and Technology (NIST), "The Smart Grid Interoperability Panel Cyber Security Working Group: Smart Grid Cybersecurity Strategy and Requirements, revision 1," accessed Aug. 2015. [Online]. Available: http://csrc.nist.gov/publications/nistir/ir7628/nistir-7628_vol2.pdf
3. C. H. Hauser, D. E. Bakken, I. Dionysiou, K. H. Gjermundrod, V. S. Irava, J. Halkey, and A. Bose, "Security, trust, and QoS in next generation control and communication for large power systems," *International Journal of Critical Infrastructures*, vol. 4, no. 1/2, 2008.
4. R. Vaswani and E. Dresselhuys, "Implementing the right network for the smart grid: Critical infrastructure determines long-term strategy," Silver Spring Networks, White Paper, accessed Aug. 2015. [Online]. Available: http://www.silverspringnet.com/pdfs/whitepapers/SilverSpring-Whitepaper-UtilityProject.pdf
5. A. Aggarwal, S. Kunta, and P. K. Verma, "A proposed communications infrastructure for the smart grid," in *Proc. Innovative Smart Grid Technologies (ISGT)*, Gaithersburg, Maryland, USA, Jan. 2010.
6. A. Abdallah and X. Shen, "Lightweight security and privacy preserving scheme for smart grid customer-side networks," *IEEE Transactions on Smart Grid*, to appear.
7. H. Li, X. Lin, H. Yang, X. Liang, R. Lu, and X. Shen, "EPPDR: an efficient privacy-preserving demand response scheme with adaptive key evolution in smart grid," *IEEE Transactions on Parallel and Distributed Systems*, vol. 25, no. 8, pp. 2053–2064, Aug. 2014.
8. M. M. Fouda, Z. M. Fadlullah, N. Kato, R. Lu, and X. Shen, "A lightweight message authentication scheme for smart grid communications," *IEEE Transactions on Smart Grid*, vol. 2, no. 4, pp. 675–685, Dec. 2011.
9. M. M. Fouda, Z. M. Fadlullah, and N. Kato, "Assessing attack threat against zigbee-based home area network for smart grid communications," in *2010 International Conference on Computer Engineering and Systems (ICCES)*, Cairo, Egypt, Nov.-Dec. 2010, pp. 245–250.
10. M. M. Fouda, Z. M. Fadlullah, N. Kato, R. Lu, , and X. Shen, "Towards a light-weight message authentication mechanism tailored for smart grid communications," in *IEEE International Workshop on Security in Computers, Networking and Communications (SCNC'11)*, Shanghai, China, Apr. 2011.
11. T. Goodspeed, "Extracting keys from second generation zigbee chips," LAS VEGAS, NV, USA, Jul. 2009.

12. S. Blake-Wilson, "Securing the smart grid," AuthenTec Embedded Security Solutions, White Paper, accessed Aug. 2015. [Online]. Available: http://www.authentec.com/white-paper.cfm

13. M. Carpenter, T. Goodspeed, B. Singletary, E. Skoudis, and J. Wright, "Advanced metering infrastructure attack methodology," InGuardians, White Paper, Jan. 2009, accessed Aug. 2015. [Online]. Available: http://inguardians.com/pubs/AMI_Attack_Methodology.pdf

14. D. R. Stingson, *Cryptography Theory and Practice*, 3rd ed. CRC Press, Nov. 2005.

15. M. Bellare and P. Rogaway, "Random oracles are practical: a paradigm for designing efficient protocols," in *Proceedings of the 1st ACM conference on Computer and communications security*, ser. CCS '93. New York, NY, USA: ACM, 1993, pp. 62–73. [Online]. Available: http://doi.acm.org/10.1145/168588.168596

16. Mathworks - matlab and simulink for technical computing. [Online]. Available: http://www.mathworks.com/

17. M. Kgwadi and T. Kunz, "Securing RDS broadcast messages for smart grid applications," in *Proceedings of the 6th International Wireless Communications and Mobile Computing Conference*, ser. IWCMC'10. New York, NY, USA: ACM, 2010, pp. 1177–1181.

18. G. Calandriello, P. Papadimitratos, J.-P. Hubaux, and A. Lioy, "Efficient and robust pseudonymous authentication in vanet," in *Proceedings of the 4th ACM international workshop on Vehicular ad hoc networks*, ser. VANET '07. New York, NY, USA: ACM, 2007, pp. 19–28. [Online]. Available: http://doi.acm.org/10.1145/1287748.1287752

19. "OpenSSL: Cryptography and SSL/TLS Toolkit," accessed Aug. 2015. [Online]. Available: http://www.openssl.org/

Chapter 9
Conclusion and Future Directions

9.1 Concluding Remarks

In recent years, research attention on the smart grid comprising distributed power generators such as micro-grids has significantly increased [1–7]. The smart grid continues to evolve to date by integrating electric power engineering technologies with communication networks, through a plethora of power instrumentation sensors and smart meters deployed between the electricity generator and consumers. Through these assorted technologies, the smart grid is expected to offer the consumers the long-cherished opportunity to communicate with their utility provider. This empowers the consumers to express their energy needs in an interactive manner with the provider, and also allows them to monitor and regulate their own energy use. However, the smart grid communication and power distribution systems are subject to many complex issues such as architecture with legacy support, varying demand response and load management, varying price of power, power loss in smaller micro-grids, electricity storage, security, and so forth. These issues lead to various decision making challenges.

The aforementioned challenges are also interesting from whichever perspective we look at them. For example, we can formulate some problems considering that the consumers should receive the incentives, while other problems may focus on the benefit of the provider. In other words, what is an opportunity for the consumers could very well be a de-motivating factor, i.e., a trade-off challenge for the utility provider, and vice versa. Such a trade-off can be depicted as an optimization problem, and could be solved with state-of-the-art optimization techniques such as game theory. Therefore, it is essential to identify the scope and challenges of the smart grid in a comprehensive fashion so that efficient delivery of sustainable, economic, and secure electricity distribution may become possible in the smart grid. In this book, we also described some relevant techniques to address these challenges and include numeric results to show how these techniques can be effective. The content of the book can be useful for readers of a broad spectrum. The initial

© The Author(s) 2015
Z.Md. Fadlullah, N. Kato, *Evolution of Smart Grids*, SpringerBriefs
in Electrical and Computer Engineering, DOI 10.1007/978-3-319-25391-6_9

chapters could be useful for interested readers with basic knowledge in electric grids and communication networks while the later chapters might offer more insight to the advanced readers who have been active in the smart grid research area.

Frankly speaking, smart grid adoption is taking its time, and is somewhat different in each country. But the fact of the matter is—smart grids are happening. With the promises smart grid offers, many challenges follow. Adequate demand response, power loss minimization, and security are three core challenges (that this book highlighted on) to be effectively addressed in smart grid design and implementation. The smart grid does not only hold the potential to improve the energy sector but also it is being considered as an important IoT (Internet of Things) application which is currently being implemented around the world. Just like the smart grid, the IoT is about inter-connecting devices at a large scale. The AMI of the smart grid continuously monitors every electricity point within a house using which the way electricity is consumed can actually be modified. This information at a wider scale (e.g., in the city level) can be used for maintaining the load balance within the energy grid assuring a high quality of service. Scaling further, with the global focus on energy management, the IoT is expected to extend the connected benefits of the smart grid beyond the distribution, automation and monitoring being done by utility operators.

9.2   Future Directions and Caveats

The rule of engaging the energy market is swiftly changing with the smart grid. For many years, energy operators considered that the demand was a statistically random process. The operators traditionally measure only the total consumption of each customer instead of its distribution over time. With state-of-the-art smart meters able to report energy use profiles over time, the operators can come up with time-dependent tariffs to give customers an incentive to lower their peak consumption. If these tariffs are transparent to the users at real-time, and more importantly if they are reasonably stable over time, the customers can schedule their consumption to flatten their usage pattern to effectively reduce their average energy bill. In practice, this will largely depend on the good will of the operators also, whether they want to give their customers more control in determining their own usage or not still remains a concern.

The micro grids based power generation from renewable sources creates random and unpredictable changes in the total power generation. This means even more dynamic adaptation of demand at real-time is needed. The change in tariffs will be much more frequent in such cases and this may be bothersome the customers to make decisions in real-time. The customers could input their preference into their smart meter beforehand (e.g., accept the tariffs within certain ranges and hours, etc.), and they could review and change the preference when they have the time. This means that it comes down to the utility operators to actively manage the regulation of energy use based on pre-decided inputs from the customers using bi-directional machine-to-machine technology.

Smart grid is an exciting technology to many, and out of excitement, the pro-smart grid camp often forgets that the communication network and home controllers of the smart grid do come with a cost. Developing and deploying the communication network, controllers, and smart grid applications at will is far from an optimal model. They need to support general-purpose machine-to-machine communications and applications to decrease the marginal cost of the smart grid use cases. A viable vision is to consider these as IoT [8, 9] applications. With such a bigger vision, the smart grid concept can be further extended to other utility services. The improvement of gas, water, and waste-water use can be achieved if the corresponding utility providers also adopt similar means. IoT can be exploited effectively to seamlessly integrate all these. The connectivity and accessibility which the IoT offers could enhance the user experience and efficiencies permitting greater interaction and control for the customers. Also, the IoT can deliver more data for manufacturers and utility providers to reduce costs through diagnostics and city-wide meter reading capabilities. Thus, IoT will become instrumental in realizing more connected, cost-effective, and smarter utility grids.

The smart grid standards are still evolving. Some countries are keener in having and advancing the technology than others. This means that political will to invest time and money into the smart grid is going to be a critical factor in determining its success. Also, the governments have to reach out to the masses by spreading consumer education on the impact and benefits of the smart grid technology, and empower them to take the right steps to protect their privacy. Besides government-backed initiatives, private sector investment into smart grid is increasing remarkably. The tech-giant Google is also showing a great deal of interest into the smart grid market by supporting renewable energy projects, data-center efficiency, and home energy management software and hardware tools [10]. Google itself is a major investor in the intermittent green energy market such as wind and solar projects around the globe. By holding a 37.5 % stake in the Atlantic Wind Connection, Google is emerging as a stakeholder in the next-generation grid projects. Its competitors like Honeywell and Apple are also following suit, particularly in building energy management technologies. This gives us a sense that smart grid is going to further evolve in future, and thus, will reshape our life for the better, and revolutionize the business models in the energy sector dramatically.

References

1. M. M. Fouda, Z. M. Fadlullah, N. Kato, R. Lu, and X. Shen, "A lightweight message authentication scheme for smart grid communications," *IEEE Transactions on Smart Grid*, vol. 2, no. 4, pp. 675–685, Dec. 2011.
2. Z. M. Fadlullah, M. M. Fouda, N. Kato, A. Takeuchi, N. Iwasaki, and Y. Nozaki, "Toward intelligent machine-to-machine communications in smart grid," *IEEE Communications Magazine*, vol. 49, no. 4, pp. 60–65, Apr. 2011.
3. H. Liang, B. J. Choi, W. Zhuang, and X. Shen, "Stability enhancement of decentralized inverter control through wireless communications in microgrids," *IEEE Transactions on Smart Grid*, vol. 4, no. 1, pp. 321–331, Mar. 2013.

4. M. M. Fouda, Z. M. Fadlullah, N. Kato, R. Lu, and X. Shen, "A lightweight message authentication scheme for smart grid communications," *Smart Grid, IEEE Transactions on,* vol. 2, no. 4, pp. 675–685, Dec. 2011.
5. H. Liang, A. Tamang, W. Zhuang, and X. Shen, "Stochastic information management in smart grid," *IEEE Communications Surveys Tutorials*, vol. 16, no. 3, pp. 1746–1770, Mar. 2014.
6. A. Abdallah and X. Shen, "Lightweight security and privacy preserving scheme for smart grid customer-side networks," *IEEE Transactions on Smart Grid*, to appear.
7. H. Li, X. Lin, H. Yang, X. Liang, R. Lu, and X. Shen, "EPPDR: an efficient privacy-preserving demand response scheme with adaptive key evolution in smart grid," *IEEE Transactions on Parallel and Distributed Systems*, vol. 25, no. 8, pp. 2053–2064, Aug. 2014.
8. Y. Kawamoto, H. Nishiyama, N. Kato, N. Yoshimura, and S. Yamamoto, "Internet of things (IoT): Present state and future prospects," *IEICE Transactions on Information and Systems*, vol. E97-D, no. 10, pp. 2568–2575, Oct. 2014.
9. Y. Kawamoto, H. Nishiyama, Z. M. Fadlullah, and N. Kato, "Effective data collection via satellite-routed sensor system (SRSS) to realize global-scaled Internet of Things," *IEEE Sensors Journal*, vol. 13, no. 10, pp. 3645–3654, Oct. 2013.
10. J. S. John, "What Is Google Plotting for the Smart Grid?" accessed Aug. 2015. [Online]. Available: http://www.greentechmedia.com/articles/read/what-is-google-plotting-for-the-smart-grid

Printed in the United States
By Bookmasters